Geography Education's Potential and the Capability Approach

"This book is a very welcome entry into the status of subject knowledge in contemporary schools not only in England but also internationally. By building its theoretical background on the 'powerful knowledge' and the capability approach, it beautifully highlights the role of subject teachers as professional curriculum makers. The book shows how geography as a school subject can develop students' capabilities and enhance their wellbeing."
—Professor Sirpa Tani, *University of Helsinki, Finland*

"This book has inspired new curriculum thinking about the broader purposes and values of geography in schools. This book expertly demonstrates how concepts of powerful knowledge, capabilities, and teacher leadership intertwine to support a geography curriculum that develops human potential and freedoms. It offers a road map to improving the relevance, appeal, and applicability of geography as a fundamental and essential subject in education."
—Dr Michael Solem, Co-Director, *National Center for Research in Geography Education, USA*

"This is an important book. The ideas in the book enable school leaders and teachers to really focus on why their subject expertise matters in education and how a focus on this can benefit the education of young people. For schools this can be significant as it ensures knowledge development is at the heart of a good school, something so often taken for granted and lost in the push for results and league table positioning."
—Andrew McCleave, Headmaster and teacher of geography, *Ballard School, UK*

Richard Bustin

Geography Education's Potential and the Capability Approach

Geocapabilities and Schools

palgrave
macmillan

Richard Bustin
Worcester Park, Surrey, UK

ISBN 978-3-030-25641-8   ISBN 978-3-030-25642-5 (eBook)
https://doi.org/10.1007/978-3-030-25642-5

© The Editor(s) (if applicable) and The Author(s) 2019
This work is subject to copyright. All rights are solely and exclusively licensed by the Publisher, whether the whole or part of the material is concerned, specifically the rights of translation, reprinting, reuse of illustrations, recitation, broadcasting, reproduction on microfilms or in any other physical way, and transmission or information storage and retrieval, electronic adaptation, computer software, or by similar or dissimilar methodology now known or hereafter developed.
The use of general descriptive names, registered names, trademarks, service marks, etc. in this publication does not imply, even in the absence of a specific statement, that such names are exempt from the relevant protective laws and regulations and therefore free for general use.
The publisher, the authors and the editors are safe to assume that the advice and information in this book are believed to be true and accurate at the date of publication. Neither the publisher nor the authors or the editors give a warranty, express or implied, with respect to the material contained herein or for any errors or omissions that may have been made. The publisher remains neutral with regard to jurisdictional claims in published maps and institutional affiliations.

Cover illustration: © ANDRZEJ WOJCICKI / Getty

This Palgrave Macmillan imprint is published by the registered company Springer Nature Switzerland AG
The registered company address is: Gewerbestrasse 11, 6330 Cham, Switzerland

Dedicated to my wife Sarah, and daughters Elizabeth and Evelyn.

Foreword

The publication of this book is a milestone and I am thrilled to have been asked by Richard Bustin to write this Foreword. Before I continue, I should declare an interest (or two). Richard was a part-time doctoral student who I had the pleasure of supervising from October 2010 to Spring 2016: he completed his PhD in that time period whilst also holding down a full-time teaching post and getting married. He was also a highly active school-teacher partner in the EU funded GeoCapabilities project[1] which I led (2013–2017). He has therefore both drawn from and contributed to a growing discussion about what we call the 'capabilities approach' to secondary school geography teaching. This is an idea which has now taken root internationally—in parts of China, Japan, Australia and the United States of America (USA), as well as several countries in Europe. The ideas informing and structuring these discussions have been gestating for quite some time—including in my own inaugural professorial lecture in 2009 and my work with John Morgan during the years we 'job shared' at the Institute of Education (IoE) (work cited frequently by Richard). Thus, it is truly wonderful to see a book-length account detailing Richard's research and his original interpretation of the capabilities approach. Written by a practitioner and someone who has been involved through much of the gestation period as an insider-researcher, teacher,

[1] www.geocapabilities.org. Comenius project (539079-LLP-1-2013-1-UK-COMENIUS-CMP).

protagonist and commentator, this book will provide a rich source for geography educationists for years to come.

To publish a doctoral thesis in book form is easier said than done, though Richard is not unique to have done so in the field of geography education research (e.g. Brooks 2016; Deen 2015). When this happens, it naturally becomes a moment for reflection, not least on the *significance* of the painstaking research, which now becomes available to a wider public readership. By coincidence, *this* moment of reflection coincided with my official retirement from University College London (UCL) at the beginning of 2019 (at least from full-on academic and educational pursuits) and with an invitation, in January 2019, to contribute to a UCL Institute of Education (IoE) debate[2] entitled "What if … our main objective in education was to build wisdom?" In considering my contribution to this debate, I was not surprisingly drawn into thinking about some of the discussions that have been part and parcel of GeoCapabilities thinking, the subject of this book. Wisdom is of course quite a slippery idea: according to Wikipedia, it involves "the ability to think and act using knowledge, experience, understanding, common sense and insight". It is thus, like the idea of capability—the precise genealogy of which is important and fully explored in this book—a broad and generic human attribute. And like capabilities, wisdom is unevenly distributed. Some folk appear to have a lot of it. Others, including sometimes highly educated public figures occupying influential positions of leadership, ostensibly seem not to be blessed with great wisdom. This is not to say that education is irrelevant of course. If we lack "knowledge, experience and understanding", we may grow to depend on 'received wisdom', which carries far less of a positive note than wisdom itself—which is correctly revered and looked up to. But building wisdom is clearly not *all* about education. Thus, wisdom is often associated with maturity and old age, stressing the role of experience and perhaps the 'university of life'.

So, wisdom is not necessarily the product of a 'good education': we can point to the most reckless and unwise decisions made by the highly educated elite. And just as easily, we can point to instances of profound wisdom emanating from 'uneducated' individuals dependent almost entirely

---

[2] https://www.ucl.ac.uk/ioe/news-and-events/ioe-public-debates.

on local funds of knowledge (González et al. 2005). However, in the context of the Anthropocene, the newly recognised 'human impact epoch' of planet Earth's geological history (Rawding 2018), which requires ways of thinking about human-environmental relations which appreciate the systematicity of interrelated physical, chemical, biological and human processes, we cannot ignore the crucial part 'schooled society' (Baker 2014) must play if human groups, including nations, are to confront the predicaments of the Anthropocene with any wisdom. The specialist knowledge required to grasp the way the Earth's systems work and interact with human activity is now, arguably, a democratic requirement of all. As I argued in the IoE debate and repeat here: there are no guarantees, but education is all we have. Furthermore, we are where we are: human ingenuity has brought humanity to the brink of causing rapid and irreversible change to the global commons, which is already impacting on human wellbeing and biodiversity (and many other such metrics) differentially everywhere. Popular political responses appearing to promise a retreat from the global and to build metaphorical (or actual) walls in order to keep reality at bay are surely inadequate to the challenges. Thus, creating educated societies that not only are able to hold leaders to account but also make the running in what constitutes some kind of collective wisdom in adapting to the dilemmas of the Anthropocene, would at least appear worth trying. If we imagine an educated public who can (amongst other things) 'think geographically', we may at least have a public which is a little more prepared for living in the human epoch than those who cannot.

This kind of argument is difficult to make. For one thing, it runs the risk of amounting to nothing more than "motherhood and apple pie": warm words, impossible to disagree with but with no real bite. And of course, the argument can easily be misinterpreted and misrepresented. Thus, when we argue for the importance of specialist, formal knowledge of the kind taught and learned in school—what Richard (after Michael Young) refers to as *powerful knowledge* in this book—it seems to some that we are being dismissive of experience and local knowledges. (Not so. It is not to hold a deficit view of someone to suggest that there is something valuable that they can be taught; quite the reverse in fact.) Worse than this, to many the argument for specialist subject knowledge in the school curriculum is dismissed as being retrogressive, a call for a traditional, Gradgrind schooling that is inappropriate for the twenty-first century. To question the role of scientific knowledge in the school curriculum, and to cast doubt on it being an essential *component* of the building of wisdom, may simply result from the contemporary educational *zeitgeist*. This continues to be heavily influenced by the OECD (Organisation for Economic Co-operation and Development) agenda for economic growth and social progress,[3] which urges national education systems towards generic skills outcomes, usually described as transversal competences.

But what such negative perceptions of knowledge-led curriculums overlook is the concern to address precisely what Michael Young and colleagues imply about the particular *qualities* possessed by powerful knowledge. In short, this draws out the distinction between the delivery of a Hirschian fact-based 'core knowledge' curriculum (Hirsch 2006) and a curriculum of engagement with knowledge in which learners are initiated into forms of *know how* (in addition to *knowing that*). This strives to give young people some insight into how knowledge gains its warrant (i.e. how do we know that such-and-such is 'the best knowledge we have') and how knowledge acquires a constant state of becoming. Such a dynamic conception of knowledge—being a human construction, it is always open to question and contest—goes to the heart of what we mean by powerful knowledge. A curriculum based on powerful knowledge will be enacted using appropriate pedagogic strategies to ensure that all learners are initiated to some degree with the procedural 'know how', of argumentation for example, in order to make good distinctions and healthy allegiances towards 'what we know'. In addition, a second form of know how (Winch 2013), of how knowledge formation necessitates building systematicity, is crucial. This takes us to the arena of 'big concepts' (or the meta-concepts) of the subject, which in geography includes place, space and environment; and to the so-called threshold concepts of the subject such as 'interdependence' or 'global'. In sum, these powerful concepts, which bring together constellations of linked processes and substantive

---

[3] "The OECD's work on education helps individuals and nations to identify and develop the knowledge and skills that drive better jobs and better lives, generate prosperity and promote social inclusion". http://www.oecd.org/education/.

concepts, indicate what is involved in thinking geographically. Thinking effectively about the dilemmas of the Anthropocene, for example, requires the geographical grasp of interdependence on a number of scales incorporating aspects from across human and physical processes.

Approaches to imagining the kind of curriculum thinking indicated in the previous paragraph are quite subtle. We are arguing for a knowledge-led curriculum, but not any old knowledge-led curriculum. Here, Young and Muller's (2010) well-known "three curriculum scenarios" paper has been enormously helpful. The Hirschian, fact-based scenario was characterised as Future 1 with an inert view of 'given' knowledge; in other words, based on a grossly under-socialised view of knowledge. This was contrasted with a grossly over-socialised view of knowledge, so-called Future 2, which is exemplified by the OECD approach, where the knowledge is considered arbitrary. All lessons have content of course, but the content selection is flexible in Future 2 thinking and can be heavily influenced by the experiences and everyday lives of the learners themselves (to be 'relevant' to them). In short, teachers are no longer concerned with *what* they should teach, but are instead focussed on facilitating the learning competence of the students. In this way Future 2 can be characterised as being an *abrogation* of curriculum responsibility, acceptable only if the sole function of schools were to prepare flexible young people for the 'twenty-first century workforce'.

However, as should be clear from the position I have outlined so far, I consider that educational and economic policy making need to be distinguished. Education has broader and deeper goals and should focus on the emancipatory power of knowledge on the individual. This is where Future 3 comes in. As Richard Bustin shows in this book, GeoCapabilities promotes Future 3 curriculum thinking based on powerful knowledge.

What the book also shows is that the aspiration of Future 3 curriculum making requires a lot from teachers. It requires deep engagement with the subject, immersion in how the subject matters can be ethically and intellectually transformed for the classroom and a constant curiosity about what constitutes 'appropriate' pedagogic practice (appropriate for the acquisition of powerful knowledge). Many will say with obvious justification that so long as the education service of England prioritises high stakes examinations for 16 year olds, the scores of which are presented as national league tables and so promote teaching to the test, there is little chance of Future 3 curriculum thinking being enacted in practice. However, Richard's book is far more optimistic than this.

As Walter Parker (2017) has shown, teachers are always having to 'swim upstream' usually against most National Curriculum reform efforts which are structured by downstream, institutional forces:

> (Teachers) must swim upstream, against structural currents that are dedicated to sculpting the school into its present form ... the legal system, the political economy, and the cultural norms and folk beliefs of families, religions, and new media.

But he notes that,

> it is likely that schools can do *something*. After all, schools have potent resources that are institutionally unique to them; namely classrooms, teachers, curriculum, instruction, assessment, materials, and students whose parents have sent them to school for the purpose of learning its curriculum. It is within this agentic space that ... 'courses and programs' are created and will have whatever effects they do. (Ibid., pp. 258–259)

I agree with that. And as Richard shows, through the capabilities approach to teaching, we have a framework for imagining the profoundly important agentive space of geography education.

Emeritus Professor of Geography Education David Lambert
at the UCL Institute of Education, London, UK January 2019

References

Baker, D. (2014). *The Schooled Society: The Educational Transformation of Global Culture*. Stanford University Press.
Brooks, C. (2016). *Teacher Subject Identity in Professional Practice: Teaching with a Professional Compass*. Abingdon: Routledge.
Deen, A. K. (2015). *Geographical Knowledge Construction and Production: Teacher and Student Perspectives*. Boca Raton, FL: Universal Publishers.

González, N., Moll, L., & Amanti, C. (2005). *Funds of Knowledge: Theorizing Practices in Households, Communities and Classrooms*. Mahwah, NJ: Lawrence Erlbaum Associates.

Hirsch, E. D. (2006). *The Knowledge Deficit: Closing the Shocking Education Gap for American Children*. New York: Houghton Mifflin.

Parker, W. C. (2017). Toward Powerful Human Rights Education in Schools. In J. A. Banks (Ed.), *Citizenship Education And Global Migration: Implications for Theory, Research, and Teaching* (pp. 457–481). Washington, DC: AERA.

Rawding, C. (2018). The Anthropocene and the Global. In M. Jones & D. Lambert (Eds.), *Debates in Geography Education* (2nd ed.). Abingdon: Routledge.

Winch, C. (2013). Curriculum Design and Epistemic Ascent. *Journal of Philosophy of Education, 47*(1), 128–146.

Young, M., & Muller, J. (2010). Three Educational Scenarios for the Future: Lessons from the Sociology of Knowledge. *European Journal of Education, 45*(1), 11–27.

Preface

I remember sitting in Professor David Lambert's office back in 2009 to talk about the possibility of doing some form of doctoral research. At that time, he was the Professor of Geography Education at the UCL Institute of Education (IoE); I was a secondary school geography teacher who had just completed the unique MA Geography Education offered at the IoE. I was working in a school with a particular set of values; the geography I was teaching was very different to that which I had encountered at University (something I chose to explore as part of my master's work), the examination authorities seemed to differ too in the sorts of geography they were offering. I wanted to explore the role, purpose and value of geography as a school subject in education. It was in this meeting that David shared with me some of the ideas he had been developing using the concept of the 'capabilities approach', which was being applied to thinking about school geography. This was the first time I encountered this set of ideas, although as a geographer I was aware of some of the outcomes of the capabilities approach to development studies. I went away from that meeting with a few papers to read and loads of ideas, and from that I did further research before putting a formal 'research proposal' in to study 'GeoCapabilities'. So began an academic adventure that involved funded research projects with trips around Europe, sharing my research and ideas internationally and finally the doctorate and book you see before you. I was honoured that Professor Lambert supervised my research, led the funded projects and that he has written the Foreword to this book.

This book is unashamedly aimed at an 'academic' audience; key writers and ideas have shaped the content of the book and their work is cited throughout. As such, it will be of interest to writers, researchers and academics in the field of geography education and curriculum studies. Given there are many ways into teaching geography, those at the start of their teaching career will find much here to reflect on, especially if the route into teaching has a limited amount of taught, theoretical content. Experienced teachers will be able to use their experience to reflect on many of the ideas presented here and as such will also benefit from the reflection the book offers. Many schools are increasingly looking to research to inform practice and as such, there is much that Senior Leaders in schools can take from the book in terms of curriculum organisation and principles. Headteachers and Deputy Headteachers in schools will be able to take ideas from this book to aid curriculum design and implementation.

Chapter 1 sets the scene, by outlining some of the main issues underpinning the ideas in this book—the differing aims of schools, the role of knowledge and skills, the role of teachers as subject specialists, and the potential of school geography through the capabilities approach. Chapters 2 and 3 unpick the first of two central ideas, the notion of 'curriculum'; why there has been talk of a curriculum 'crisis', and geography, as an academic discipline and a school subject and the relationship between the two. Chapter 4 is about the second of the two central ideas, the capabilities approach, and how its ideals and values have transcended its original formulation in welfare economics and are being used more in education. It is in this chapter that ideas are brought together to explain the formulation of GeoCapabilities. Chapter 5 focuses on the ways in which the notion of GeoCapabilities has been developed through research, two international research projects, doctoral research and various workshops with teachers and academics in conferences and schools throughout the world. Chapter 6 offers a vision for both a school and a geography curriculum that is built on the principles developed throughout the book; a critique of the ideas is offered as well as a series of challenges that exist to meeting this vision in practice.

Throughout the book, the ideas and terms are based in an English schools setting, yet it was during the research projects that are explained in Chap. 5 that an international team of geography educationalists saw the possibilities of the capabilities approach to curriculum thinking for their countries. So the ideas of GeoCapabilities and powerful knowledge transcend the English setting and can be interpreted for curriculum thinking around the world, offering an international potential for geography education. Chapter 1 briefly introduces the nature of geography education in the USA and Finland as a comparison with the English experience.

A key inspiration for the book and for the journey of my research is the story of my great-grandfather, Mr Alfred Bray Treloar. He was headmaster of Tavistock Grammar School, in Devon, from 1920 to 1946. I have always felt a connection to Mr Treloar as in a book written to celebrate the centenary of the school in 1978, *Tavistock Grammar School: The first thousand years* (Woodcock 1978) we are given a glimpse into a school trip he ran in 1931:

> On March 23rd 1931 he took a party of boys to see a football match between Plymouth Argyle and Nottingham Forest. If the boys were expecting a carefree afternoon they were soon disillusioned, for they discovered that the main purpose of the outing was for them to write notes under the following headings:
>
> (a) Features of interest en route,
> (b) The China Clay mineral track at Marsh Mills,
> (c) The Plym Estuary,
> (d) A holiday crowd and typical scenes attending such occasions,
> (e) The game as it should be played. (Woodcock 1978, p. 145)

I have often been struck by this recount of a school trip. Many of the themes the pupils are reflecting on are geographical in nature and would not be out of place in contemporary geography lessons—features of interest seen in a journey, location and features of local industry (china clay), formation of an estuary and tourism geography. This in a sense was a 'field trip', albeit tied up with a football match, and Mr Treloar was encouraging his pupils to observe and to understand the landscape around them to help them make sense of the world. When I run field trips of my own now, I encourage my pupils to look out of the window as we travel to our destination and one of my first questions when we arrive is always about what we saw out of the window and how the landscape changed.

Much has changed since Mr Treloar's field trip of 1931. Tavistock Grammar School no longer exists, and Tavistock itself is a bustling tourist town, its china clay workings now a part of its industrial history. His 1931 classroom is a world away from the modern classroom. No computers, iPads, laptops, interactive white boards or Geographical Information Systems.

Yet I still feel Mr Treloar understood that there was something uniquely important about getting the pupils out of the classroom, and about getting them to reflect on their surroundings, to engage with local industry, tourism and physical landscapes. Although the methods may have changed, and the technology available to teachers different, it seems to me that geographical knowledge has always had a central role to play in schooling. As the pages of this book describe, this is not how some contemporary educators see the role of schools or the organisation of a curriculum.

My great-grandfather has a plaque in his honour in Tavistock Parish Church, with the inscription "they who understand shall instruct many". It is this sentiment which underlies much of my professional practice as a geography teacher but also, more recently, my understanding of the changing educational landscape and my belief in the significant role that subjects such as geography play in schools. This is why I talk of the 'potential' of a geography education in a curriculum.

The concept of 'GeoCapabilities' that is used in this book to frame discussions around the value of a geography curriculum has been increasingly appearing in geography education. Developed over a series of papers published across the world, it was advanced through workshops, lectures and projects; yet, to date, there has been no attempt to pull all this thinking together coherently. That is what this book sets out to do; to tell the story of the development of the concept so far and the ideas behind it. Yet in the process of writing the book, more questions were created than answered. So this book should certainly not be seen as the last word on 'GeoCapabilities'; in fact, the conversations it engenders are only just beginning.

Worcester Park, Surrey, UK   Richard Bustin

Reference

Woodcock, G. (1978). *Tavistock School, the First Thousand Years*. Billing and Sons, Ltd.

Acknowledgements

Thanks to the teachers I have worked with over the years, and students past and present who are my constant source of inspiration. Thanks too to my colleagues on the GeoCapabilities project, Professor Lambert and colleagues at the UCL Institute of Education, and friends and colleagues at the Geographical Association. I would also like to thank my own school geography teachers - Mr Seymour, Mr Cowx, Mr Hatch and Mr Pafford for inspiring a lifelong love of geography.

Contents

1 What Is the Purpose of Schools?   1

2 Mapping a Curriculum 'Crisis'   33

3 Bringing the 'Geography' Back in   67

4 The 'Capabilities Approach' to Geography Education   99

5 Developing GeoCapabilities: The Role of Research   131

6 The Potential of a Future 3 'Capabilities' Curriculum   159

Index   193

List of Figures

Fig. 1.1 Extract from the 2013 National Curriculum showing the slimmed down content requirements for geography (DFE 2013)   18

Fig. 2.1 A model of curriculum linking content, methodologies and education to inform curriculum theory, which directly informs practice. Redrawn from Lovat (1988, p. 212, www.tandfonline.com)   35

Fig. 3.1 A timeline of the changing nature of geography as an academic discipline and school subject (based on Walford 2000; Boardman and McPartland 1993a, b, c, d)   75

Fig. 3.2 The nature of curriculum framing. The weakly framed curriculum illustrated here sees geographical knowledge as part of a 'Humanities' subject, in which it loses its geographical identity   81

Fig. 3.3 Diagrammatic representation of vertical (left) and horizontal (right) knowledge structures (after Bernstein 1996)   84

Fig. 3.4 A model of 'curriculum making in geography' (based on Lambert and Morgan 2010, p. 50)   90

Fig. 4.1 A conceptual model to show the various aspects of the capabilities approach   102

Fig. 4.2 A conceptual model of how the capabilities approach has been applied to educational discourse   112

Fig. 4.3 GeoCapabilities as expressed by Solem et al. (2013), in which powerful knowledge (as expressed by Lambert and Morgan 2010) becomes the bridge to connect the curriculum to capabilities development 125
Fig. 5.1 The results of the initial discussions about what to include in a lesson sequence on 'Russia' (from Bustin et al. 2017) 144
Fig. 5.2 The completed GeoCapabilities Framework, created for the research (from Bustin et al. 2017) 145
Fig. 6.1 A model of the capability approach to education 161

List of Tables

Table 2.1	A range of different curriculum ideologies that influenced curriculum debates in the late twentieth century (Rawling 2000)	43
Table 2.2	Curriculum ideologies and their impact in the classroom (Bustin 2018)	45
Table 2.3	The key features of 'powerful knowledge', based on Young (2008)	59
Table 2.4	Key features of powerful pedagogy (Roberts 2013b)	61
Table 3.1	The key concepts of geography (Taylor 2009)	71
Table 3.2	The relationship between classification and framing of the geography school curriculum based on Bernstein (1973) and Daniels (1987)	82
Table 3.3	A 'vignette' of powerful geographical knowledge based on the teaching of coastal geomorphology in the geography classroom	86
Table 3.4	Types of powerful knowledge in geography (Maude 2016)	87
Table 4.1	A collection of selected 'universal capabilities', based on Alkire (2002)	104
Table 4.2	A list of suggested universal 'educational' capabilities	117
Table 4.3	A list of capabilities derived from the study of humanities in higher education (Hinchliffe 2006)	119
Table 4.4	A suggested list of the capabilities approach to geography: 'Three GeoCapabilities' from Solem et al. (2013, p. 221)	121

Table 4.5	Expressions of the powerful knowledge of geography on which GeoCapabilities is based, with reference to Maude's types of geographical knowledge (2016) (Lambert 2017)	122
Table 5.1	Outcomes of the GeoCapabilities 1 project from Solem et al. (2013)	135
Table 5.2	Topics of conversation in the semi-structured interview process for two groups of interviewees (from Bustin 2017)	142
Table 5.3	The contentions derived from doctoral research into GeoCapabilities	146
Table 6.1	School teachers exploring the powerful knowledge of their subjects	168
Table 6.2	Two three part lessons from the medium term plan of a sequence on tuna fishing, part of a larger topic on natural resources (from Bustin 2015)	170
Table 6.3	The use of the practical GeoCapabilities Framework to map out the geographical knowledge component of the two-lesson sequence on tuna fishing	172

1

What Is the Purpose of Schools?

Introduction

There has never been a more important time to learn geography. Young people today are growing up in a world of climate warming bringing untold changes to the natural world including sea level rise, extreme weather events and changes to vegetation and animal species. Increased population and globalisation are moving people, money, ideas and cultures around the world creating conflict and complex geopolitical relationships. Young people also have unprecedented access to information; facts and figures can be dredged up from the depths of the internet to support or refute a myriad of claims.

Schools enable young people to make sense of the complex world around them, and place facts and figures into a broader framework of understanding. They help young people to find their place in a complex and ever changing world. Teachers are the key to enabling young people to learn geography. Using their knowledge and skills, they help challenge and question, engaging students in rigorous ideas.

Educational sociologist David Baker (2014) argues that mass education on a global scale has led to a 'schooled society' over the past 150 years, the significance of which pervades every area of political, cultural and economic life. Yet despite this 'quiet' educational revolution, more recent reports (e.g. Carson 2019) suggest that social media has contributed to the increasing acceptance of 'fake news' and false knowledge claims, also impacting cultural and political norms.

The nature of a school curriculum that enables a high-quality education is changing. Views of what constitutes knowledge and how it differs from the sorts of readily available facts is contested. The balance of a school curriculum between teaching subject knowledge and developing more generic skills and competencies seems to be in constant flux. The way teachers are trained is also changing, which has the potential to influence the nature of the very teachers standing up in front of classes of students.

This book is about what we choose to teach young people in schools, why we choose to teach it and the deep thinking that teachers and school leaders do every day to ensure our young people receive a world class education. The book is focussed throughout on the notion of 'curriculum', a term that describes what goes on in school, what is taught, why and how learning is structured. This book also discusses knowledge, and the importance of subject knowledge in the school curriculum and how this relates to academic disciplines in universities. This is a topic of much debate in recent educational discourse (e.g. Young 2008). In 'what are schools for?' Standish and Cuthbert (2018) argue:

> Many young people entering the teaching profession are unclear about the role of disciplines and knowledge in the school curriculum... For those already working in the profession, including experienced teachers... subjects have come to be viewed... as a means to another end such as developing marketable skills, facilitating well-being, promoting diversity or addressing global issues. (pxvii)

This suggests that the place and value given to knowledge, subjects and some of the broader aims of a curriculum are confused, and this book explores some reasons behind why this situation has arisen, the positions adopted by various thinkers who promote alternative types of curriculum and a possible framework for thinking about the curriculum that could offer some clarity on the issues.

These issues are particularly important at the end of the 2010s. In England, the curriculum has been dominated for decades by the National Curriculum prescribing content through traditional subjects, effectively removing the need for teachers to engage in curriculum thinking of their own. As more schools are developed that are outside the influence of the National Curriculum, so the need to understand curriculum becomes more pertinent. In 2016 and 2017 reforms of examination criteria for students sitting exams at the age of 16 (General Certificate of Secondary Education, GCSEs) and 18 (A-Levels) created a content rich curriculum. Yet the types of knowledge being promoted here seemed to place value on recitation of learnt facts, rather than seeing any role for knowledge in developing further understanding or in any way being enabling for young people. It is this latter aspiration for knowledge that has led to talk of 'powerful knowledge', a term from Michael Young (2008) whose ideas are central in this book. It is a curriculum built on the powerful knowledge of subjects that invests subjects with their educational potential.

Framing many of the debates about powerful knowledge, subjects and the curriculum in this book is the 'capabilities approach' (e.g. Sen 1980; Nussbaum 2000), applied to educational discourse. This provides a means to consider what a curriculum is able to enable a student 'to be' or 'to think like' as a result of their education. As such, it provides a holistic view on the aims of education expressed in terms of human freedoms, and the choices that an education leads to. As a framework for curriculum thinking it is more ambitious than simply seeing education in terms of passing exams, being able to recite facts, or having a set range of skills.

This book tells this curriculum story through the school subject of geography, a subject that has been described as 'one of humanities' big ideas' (Bonnett 2008), yet the teaching of which has been branded 'boring' and 'irrelevant' in the past by the British Government's Office for Standards in Education (Ofsted 2011). Geography is the subject that helps young people make sense of an ever changing world, with climate change and globalisation providing a range of possible global futures, yet the subject has been accused of being 'corrupted' (e.g. Standish 2007, 2009) by 'good causes'. It is the powerful knowledge of geography, free from the corrupting influences of various good causes, that has the ability for pupils to develop 'capabilities' through their education. When applied to the geography curriculum these are called 'GeoCapabilities' (e.g. Lambert 2011a), and it is this that invests school geography with its educational potential. A potential to help young people discern fact from fiction; a potential to understand the nature of the changing world; and a potential to be able to think knowledgeably, live and work in the modern world.

This book sets out to explore these ideas.

Setting the Scene: In Search of a Curriculum 'Big Picture'

Most schools have a set of 'aims' proudly displayed on school websites and promotional materials. Yet if the curriculum could be redesigned from scratch to meet those aims, it would be interesting to see how many schools would still end up with a rigid subject-based curriculum and not one based around skills and competencies. Subjects as the basis of a curriculum seem to have always existed, and in schools children go from subject to subject over the course of a day without any sense of an overarching bigger picture explaining why they do this, or what the connection is between these subjects and some overarching greater aim. A number of possible constructs can be used to develop a 'big picture' of the curriculum to provide this overarching set of ideals, not least the National Curriculum itself.

The National Curriculum

With the introduction of the National Curriculum in 1988 into all state run schools in England, the content of the school curriculum was 'fixed' into a series of ten subjects; English, mathematics, science, technology, history, geography, a modern foreign language, music, art and physical education (PE). These subjects were not new creations, but had been the basis of schooling since the turn of the century. As John White (2006) asserts "the 1988 curriculum could almost have been lifted from the 1904 regulations for the newly created state secondary schools" (p. 2). Each subject had a specified list of content, and the role of the teacher was to deliver and assess this content to children in classrooms. The teacher's professional role since 1988 had been to identify ways of delivering this prescribed content in dynamic and engaging ways, rather than worrying about grander ideas about aims, values and knowledge content. Yet there was no overall set of aims driving the creation of the National Curriculum. As White (2006) claims:

> [W]hen the National Curriculum appeared in 1988, it was all but aimless. It consisted of a range of subjects, but lacked any account of what these subjects were for. (p. 1)

Aims were not created until 1999, but this was problematic as "the aims came after the laying down of the subjects. Almost all of these subjects had been compulsory since 1988 and dominant for decades before this" (Ibid., p. 4). The aims of education were therefore imposed on a curriculum structure that already existed.

By 2008, after a decade of rule by a modernising Labour government, the National Curriculum 'big picture' (QCA 2008) appeared, a model of the school curriculum. A clear aim of education was articulated, which was to create "successful learners, confident individuals, and responsible citizens"; the rest of the curriculum was structured to meet these aims. The 'big picture', whilst ambitious, showed a lack of emphasis on subjects and knowledge. Subjects appear in a small and diminished role as part of 'statutory expectations'. The aims reduced education to a set of ideals, which Ledda (2007) asserted "are worse than irrelevant. They are anti educational" (p. 15). These aims made no mention of knowledge, or of subjects. The three aims could equally apply as aims of good parenting, rather than aims of a national education system. The school curriculum was seemingly influenced by the 'learning power' philosophy, summed up by this quotation from the 'campaign for learning' (Holt 2015):

> Since we cannot know what knowledge will be most needed in the future, it is senseless to try to teach it in advance. Instead, we should try to turn out people who love learning so much and learn so well that they will be able to learn whatever needs to be learned. (Holt 2015)

The view suggests that subject knowledge is irrelevant, archaic and no longer suitable for young people. It suggests that the only reason knowledge is taught to young people is in case it is 'needed in the future'. This perhaps explains why personal, social and health education (PSHE) and citizenship were introduced as new subjects into the National Curriculum; knowledge in these subjects has perhaps a greater direct relevancy to everyday life than a Shakespeare poem, or understanding the reasons behind the Second World War. This more 'traditional' knowledge had seemingly become defunct and outdated. With a 'love of learning' young people should be able to discover these things for themselves if they want to in the future.

Yet there is a problem with this assumption. Firstly, it assumes young people will grow up to only ever be passive recipients of knowledge, believing whatever anyone tells them, rather than being involved in the generation or critique of what they are being told. The ability to challenge and question knowledge, and understand how ideas have developed over time, requires the disciplined thought processes offered by subjects. But pro-subject, pro-knowledge arguments were considered 'traditional' and outdated at the start of the 2010s. The view of a pro-skills and pro-vocational education, dubbed the 'progressive' arguments, were gaining prevalence.

A decade later and the curriculum pendulum had swung the other way, creating the 'knowledge rich' curriculum of the 2010s. Gone was the model of the 'big picture' with its supposedly anti-educational ideals and in its place a definitive statement about knowledge in schools:

> The national curriculum provides pupils with an introduction to the essential knowledge they need to be educated citizens. It introduces pupils to the best that has been thought and said, and helps engender an appreciation of human creativity and achievement. (Department for Education 2014, accessed 2018)

The idea of 'essential knowledge' reasserts the importance of knowledge through traditional academic subjects, yet it opens a wider debate about what knowledge can be considered 'essential' and what makes it essential. A focus on 'knowledge' was back on the National Curriculum for schools, as a response to the skills agenda of the previous decade. Each subject community sought to create a list of 'core knowledge' of what children should be learning in schools in that subject. Michael Gove, the then education secretary, along with his education minister Nick Gibb held a view of knowledge in education that was similar to that espoused by Hirsch (1988), who wrote a book detailing 'what every American needs to know'. This is lists of facts and ideas deemed to be appropriate for each age of child. But knowledge in this tradition is static, uncritical and assumes that each subject does have a 'list of content' than can easily be drawn and created and passed down 'from generation to generation'. It was akin to the early versions of the National Curriculum with its detailed prescriptive content that simply needed to be learnt.

The 'knowledge turn' (as Lambert 2011b called it) influenced much of the educational policy in the second half of the 2010s, and it was not just at key stage 3 where changes were afoot. Rewrites of the GCSE and A Level courses occurred in 2015 and 2016. The content of the new A Level courses was influenced by the newly created ALCABS, 'A Level content advisory boards' made up of a seemingly random selection of academics from various fields keen to ensure their research interest was part of the new A Level courses. These new courses were very detailed on content to be learnt.

The 'knowledge turn' of the National Curriculum seemed to result in children learning page after page of facts with little critical engagement or reflection on what was being learnt; and not a large opportunity for children to develop their own values as a result of the knowledge gained.

The Classical Trivium

Other attempts have been made to provide a coherence for a seemingly fragmented curriculum. In 'Trivium 21C: Preparing young people for the future with lessons from the past', Martin Robinson (2013) uses a curriculum structure that was created and developed from classical times through to the Middle Ages. This three-part structure of curriculum includes *Grammar*, what might be considered the 'essential' or 'core' knowledge; *Dialectic*, the ability to question and reason through knowledge to create further understanding; and *Rhetoric*, to be able to communicate ideas and develop flair and confidence. He argues that a re-imagining of these concepts for the twenty-first century could provide a means to see beyond some of the curricular issues around knowledge, skills and an overemphasis on examination grades and league table positioning. In many ways, the aspirations of Robinson (2013) are echoed in the work of capabilities, providing curricular coherence and a grand vision for schools. Yet there is still a fundamental question over the nature of knowledge, what constitutes this 'core' or 'essential' knowledge and how decisions about that are derived.

The Habits of Mind

Another possible curriculum structure that attempts to unify the school experience is the 'Habits of Mind', based on the work of Costa and Kallik (as discussed in the Habits of Mind Institute 2018), which is explained for secondary teachers in a book by Boyes and Watts (2009). As the Habits of Mind website explains:

> [T]he Habits of Mind give learners of all ages and at all stages, a framework for autonomous, lifelong learning. They show us how to behave intelligently, independently and reflectively. (HoM Institute 2018)

The habits are broad learning competencies that offer a way for young people to learn a range of diverse skills through the subject curriculum. The 16 Habits include:

> Persisting; Thinking Flexibly; Thinking about Thinking (Metacognition); Striving for Accuracy; Questioning and Posing Problems; Applying past Knowledge to New Situations; Gathering Data through All Senses; Responding with Wonderment and Awe; Taking Responsible Risks; Finding Humour; Thinking Interdependently. The Habits of Mind (from the HoM Institute)

These various 'habits' are designed to be developed by pupils whilst in schools but what is less clear is the mechanism through which these

Habits should be developed. If a curriculum was built around these learning habits, it is unlikely a subject-based curriculum would result. Subjects would get in the way; they become an inconvenience. So whilst the Habits do refer to subjects briefly, they support a curricular model devoid of 'essential' knowledge. They are built around a set of generic competencies.

Setting the Scene: The Place of Subject Knowledge in the Curriculum

There seems to be a fundamental problem with the purpose of schools. On the one hand, there is a need to enable children to develop a range of valuable attitudes and beliefs befitting a twenty-first century child whilst at the same time enabling them to develop knowledge and skills valuable for the modern post-industrial workplace. This position has been summarised by Michael Young (2003), who argues:

> [A] growing tension has become apparent between the fluidity and openness to innovation of successful advanced economies—what some have termed 'fast capitalism'—and the persistence of relatively rigid divisions between the different school subjects and disciplines and between curriculum knowledge in general, and the knowledge that people use in employment and more generally in their adult lives. (pp. 99–100)

For Young (2003), the tension is between the rigidity of subject knowledge and the sort of knowledge children will be engaging with in their adult lives. Subject knowledge has been deemed old fashioned and 'traditional' and subjects have taken a diminishing role in some schools as the curriculum is built around competencies.

It was due to a lack of overall vision for education that subsequent re-writes of the National Curriculum from 1988 onwards changed the nature and importance placed on subjects and knowledge. New subjects were added to fill a perceived inadequacy in a school's educational provision. 'Personal, social and health education' (PSHE) was introduced in 2000 to teach children the importance of healthy lifestyles, sex education, and relationships; 'citizenship' was introduced in 2002 to teach children the importance of voting, and what it means to be part of British society. As QCA (1998) argued at the time:

> [C]itizenship education is urgently needed ... if we are to avoid a further decline in the quality of our public life and if we are to prepare all young people for informed participation, not only in a more open United Kingdom, but also in Europe and the wider world, as we move into the next century. (p. 14)

These new subjects were set up to provide children with the 'essential' knowledge that was deemed not to be on offer elsewhere in a traditional subject-based curriculum. Yet on closer inspection these new subjects were simply amalgamations of bits of existing subjects—much of PSHE could be covered in biology and English literature; citizenship in history, geography and religious education (RE). Academic subject teachers seemed to be increasingly using curriculum time to not teach their subject; in many schools time allocated to the traditional subjects was reduced to make way for these new subjects, which meant children learnt less geography, less history, less science.

In some schools geography, history and RE were combined to form 'humanities' for 11–14 year olds. This does free up curriculum time for other 'subjects', but does mean that if a humanity teacher was a historian, and a young person chose not to study that subject beyond the age of 14, this could result in a generation of children not being taught any geography or RE in their school career.

Alongside a reduction in traditional subjects, formal education at the end of the first decade of the 2000s was becoming much more child centred and personalised, with teachers expected to know and respond to each child's preferred learning style, and to differentiate their learning activities accordingly. Emphasis for teachers was all about *how* to teach and not *what* to teach. Classrooms were busy places full of thinking skills activities and children were taking part in what Lambert (2005) has described as a 'pedagogic adventure' where children do all manner of engaging activities without a deep reflection on the knowledge they were learning. These educational ideals, as Lambert (2008) argues, have become the new orthodoxy, buoyed up with the beguiling rhetoric of 'learning to learn' and 'personalisation' but impoverishing the language of education to such a degree that I fear we may have lost track of its moral purpose. (p. 209)

This lack of a moral purpose of education is a sentiment that Furedi (2004) had noticed pervading many areas of social policy and modern society, arguing we had created a "therapy culture" in which everything is about creating well adjusted, happy young people but at the expense of any deeper meaning or knowledge development.

This lack of a moral purpose to education impacted on the nature of a subject-based curriculum. Many subjects, though particularly the humanities of geography, history and RE, enable children to engage with values. Through subjects, with good subject specialist teachers, children are able to understand climate change, different political opinions, varying religious ideals and through these subjects can arrive at their own understanding, which in turn will influence how these young people live and behave in the future. Taught badly by untrained or inexperienced teachers, values can be taught directly to children without the children understanding the issues behind the values. A similar criticism could be levelled at 'citizenship'. Children are expected to engage with 'citizenship', and 'British values' with an aim to, for example, voting in national elections without understanding the need for voting, or the historical fights that have existed in the past which have ensured our current freedom and right to hold democratic elections. As Morgan (2008) has argued, "since 1988 the work of... teachers has become increasingly tied to the needs of the economy and operated through the mechanisms of the state" (p. 20). As CIVITAS (2007) continued, "Teachers are expected to help to achieve the government's social goals instead of imparting a body of academic knowledge to their students" (p. 1). Through the notion of a 'body of academic knowledge', CIVITAS recognised the importance of academic, subject knowledge. Standish (2009) continues:

> [R]ather than teaching pupils about the world so that they can decide the most appropriate course of action, global citizenship education is tied to the specific non-academic values that tend towards the replacement of knowledge with morality as the central focus of the curriculum. (p. 39)

Knowledge was slowly being taken out of the school curriculum and was being replaced with a set of predetermined national values. As Furedi (2007) argued:

> [E]veryone with a fashionable cause wants a piece of the curriculum… increasingly the curriculum is regarded as a vehicle for promoting political objectives… (and) transmitting the latest fashionable cause or value. (Furedi 2007, pp. 1–2)

Without knowledge underlying an issue, children would be unable to query or question the nature of the values they were being promoted, nor given the chance to understand the importance or relevance of these values.

It was in part the criticism levelled at this form of curricular vision that led to a curriculum built around 'essential knowledge' yet even this sees a narrow view of knowledge. Knowledge in this vision is akin to facts, and the more facts that can be learnt the more knowledgeable a young person becomes. It says nothing about the abilities of that young person to be able to apply that knowledge, to develop understanding or to use knowledge to take a stance on moral values and issues.

Setting the Scene: The Challenge of Specialist Teaching

If a subject-based, knowledge-led curriculum is something of value in education, and this book sets out to provide reasons why this is, then well-trained, subject specialist teachers are required in front of every class to enable this. It is this that ensures young people have access to the sorts of 'powerful knowledge' that can realise the potential of a subject-based education. A number of challenges exist that is making this a difficult target to achieve.

Firstly, there has been an increase in non-specialist teachers, a product in part of the well-publicised funding crisis in schools in the second half of the 2010s. According to Kreston UK (2018), 80% of secondary academies were in deficit in 2016/2017. The result of this, according to the Association of School and College Leavers (2016), is an increase in the use of non-specialist teachers to teach classes outside their areas of expertise; their report identified 73% of the 817 school leaders surveyed resorted to this measure in response to a lack of funding.

Alongside this is a well-publicised shortage of new teachers entering the profession (e.g. Coughlan 2018). The result has been larger class sizes and increased workload for those already working in schools. The situation has been made worse by the rise in pupils studying EBacc (English Baccalaureate) subjects, which requires more teachers in those subjects. To address this, potential new teachers of physical education (PE), a subject that has not seen shortages, have been offered places on PE training courses as long as they have an A Level in an EBacc subject such as geography. Through their PE teacher training, they would be expected to teach some lessons in another (EBacc) subject. Trainees would need to study their second subject for a limited period, one provider offering a three day course (see Holbrey 2019 for a discussion of PE teachers training to teach geography), but would not be required to sit any form of assessment. Many of these specialist PE teachers have ended up in classrooms teaching subjects outside their specialism because there was a shortage of specialist teachers. A report in 'Schools Week' (Staufenberg 2018) quoted a delegate at the National Association of School Based Teacher Trainers conference who argued this plan was "effectively what schools are doing… (and it) is a bit iffy" (Staufenberg 2018). Dr Mary Bousted, the joint leader of the National Education Union, was even more critical, arguing:

[T]his speaks of government desperation; this idea that teachers can teach any subject, it's just not true. (Reported in Staufenberg 2018)

Another challenge to meet specialist teaching is time given to specialist subjects. Given the curriculum pressures on traditional subjects mentioned earlier, the time allotted to subjects has been reduced in many schools. Data from the Department for Education (2016a) identified a reduction in the hours of specialist geography teaching from 87.1% of all curriculum hours taught by specialist teachers to 83.3% at key stage 3 (11–14 year olds) and from 96.2% to 94.5% at key stage 4 (14–16 year olds) between 2010 and 2016.

The 'knowledge turn' in education was happening alongside changes to the way teachers were being trained (see Tapsfield 2016 for a discussion related to geography teachers), a product in part of the teacher recruitment crisis. The changes saw a reduction in the numbers of university-based initial teacher training courses and a shift towards more school-based training. By 2017–2018, 53% of new teachers were being trained by school-led methods, compared to 47% by higher education institutions (DfE 2017). In the university-based 'Post Graduate Certificate of Education' (PGCE) courses, groups of subject specialists train together under the supervision of university- and school-based mentors. This means beginning teachers are able to learn theory and practice and reflect on this within subject groupings. Conversely, school-based training reduces the amount of theory being taught by subject education experts, with an increase in more practical training. More significantly, through school-led routes, however, new recruits are often trained individually in schools. As Tapsfield (2016) observed:

> [T]he single trainee model, common in many school led routes, limits opportunities for trainee… teachers to work together and share best practice. (p. 108)

In schools, there may be only one new teacher in a given subject, which means they cannot work together with other new teachers in that subject. The extent to which teachers trained in this way can really develop a strong, subject-based professional identity is therefore questionable. Ironically, just as schools needed subject specialist teachers to fulfil the needs of the 'knowledge turn', opportunities to develop high-quality subject specialist teachers, as opposed to more 'generically' trained teachers, reduced.

The professional role of teachers has also been questioned. At a conference of the National Union of Teachers in 2018 (reported in George 2018), a conference delegate reported her frustrations about seeing her role as helping pupils to pass exams rather than fostering a love of a subject. As the delegate explained:

> Classrooms are now wholly about passing the exams. Students are drilled into how to do the exam and on assessment techniques. They are then tested again and again and again until they meet their over-inflated target grades. (George 2018)

This teaching to the test, a product perhaps of a system that ranks the success of schools in league tables based solely on reportable exam results, has had a crushing effect on the curriculum. As she continued in relation to the teaching of English in schools:

> Year 7 (11 to 12 year old) students… no longer write poetry in their lessons because you don't need to write poetry in GCSE any more, and students… do Macbeth every year from 7 to 11 because that is what is going to come up in their GCSE exams. (George 2018)

Teachers' professional identity is challenged in a system that values the achieving of examination grades over any other aim of schooling. It seems teaching is no longer about helping pupils discover the 'beauty' of subjects and the significance of knowledge. Teachers are the key to helping pupils realise their full academic potential and this in itself is much more than simply achieving examination grades.

Setting the Scene: The Changing Place of Geography in Schools

Some of the challenges outlined already in this chapter have suggested that the place of subjects in schools is challenged, the role of subject specialist teachers is varied and the value placed on subject knowledge has changed and continues to change over time. These debates can be related specifically to the teaching of geography in schools, and this is taken up in more detail in Chap. 3 of this book. By way of introduction to these ideas, this section looks at the experiences of geography education around the world; children growing up and going to schools in different countries have a varied exposure to a high-quality geography education.

England

'Geography' exists as a discrete subject in schools in England for children up until the end of key stage 3, at age 14, whereafter it becomes an optional subject which students can continue to study through GCSEs (with the final examination taking place at age 16), to A-Levels (taken at 18) and a myriad of degree level programmes at top universities throughout the country.

There are a number of specific challenges that the subject in schools is facing at the end of the 2010s that threaten the status of geography as a discrete subject. These include a varied engagement with the academic discipline; the increasing political curricular influence; the changing status of knowledge in the curriculum; and lack of geographical qualifications and experience of some teachers, itself a product of the changing nature of teacher training alluded to earlier.

There is an increasing number of non-specialist geography teachers teaching geography classes (e.g. Tapsfield 2016). If the teacher does not have a rigorous understanding of the geography, then the ability to introduce students to and enthuse them in the subject becomes impossible. This also coincides with an observation from Roberts (2010) that trainee teachers are not really being assessed on the knowledge content of their lessons. In 2010, only 3 out of 33 standards required for qualified teacher status actually referred to subject knowledge. As Roberts observed:

> I have become particularly concerned about the extent to which lesson plans, lessons and debriefing give more attention to general aspects of lessons than to the geography being taught and learned. (p. 112)

Teachers can be judged 'outstanding' with only a cursory mentioning of the knowledge content of geography, or worse still, with knowledge that is factually incorrect, particularly if the lesson was being observed by a non-subject specialist. Lambert (2016) observed of Roberts' (2010) writing "that a leading commentator on geography education should need to make such an obvious point is remarkable" (p. 395). Geography should be at the heart of a good geography lesson but in practice this seems to be not so.

The Office for Standards in Education, (Ofsted), the government appointed schools inspectorate, concluded in 2008 and 2011 that geography was "boring" and "irrelevant" to many of the pupils in schools (Ofsted 2008, 2011). At a national level, numbers of students opting to study the subject after the age of 14 have fluctuated; after a steady decline

1 What Is the Purpose of Schools?

from 1997 to 2007 (Butt 2008), numbers have increased in the past decade, (JCQ 2016), a product, perhaps, of the introduction of the English Baccalaureate (EBacc, DFE 2016b), a performance measure of schools which rewards children who choose either history or geography at GCSE (along with a suite of other supposedly more rigorous subjects). In spite of more children studying the subject, the quality of the geography they were learning, knowing who was responsible for teaching it, and what would happen to the future of the subject in schools were still questionable.

Throughout National Curriculum re-writes of 2007 and 2013 the prescribed content of geography reduced, with the 2013 version being printed on two sides of A4 and with a series of topic statements to cover the whole of key stage 3. Figure 1.1 is the knowledge component of the 2013 National Curriculum requirements for geography (DFE 2013).

Geography teachers were now responsible for interpreting the themes above into meaningful geography lessons. After two decades of being told what to teach and only having to decide on how best to do it, a generation of teachers were now having to innovate and decide on knowledge content for themselves. It was Morgan and Lambert (2005) who argued that lesson planning was an 'intellectual activity', not just a technical activity but doing this 'intellectual activity' successfully requires an understanding of the overall aims of geography education. The interpretation of, for example, "the use of natural resources" requires teachers to understand what aspects of natural resources they want to teach and the ideological basis for these decisions. They need to decide what sort of knowledge and understanding they want their children to gain from a course on natural resources.

The problem is teachers could teach 'the use of natural resources' or any of these topics without actually engaging with any geographical knowledge, but instead use the lessons as a vehicle for promoting another of the school or government agendas, for example, healthy eating. If pressed, a geography teacher could argue that 'food' is a resource and so healthy eating was a legitimate way to spend geography lesson time. This would arise out of a misunderstanding of the potential of geography as a school subject. There is a strong argument that needs to be made about the role of geography in schools that would help teachers to understand

> Pupils should be taught to:
>
> **Locational knowledge**
>
> • extend their locational knowledge and deepen their spatial awareness of the world's countries using maps of the world to focus on Africa, Russia, Asia (including China and India), and the Middle East, focusing on their environmental regions, including polar and hot deserts, key physical and human characteristics, countries and major cities
>
> **Place Knowledge**
>
> • understand geographical similarities, differences and links between places through the study of human and physical geography of a region within Africa, and of a region within Asia
>
> **Human and physical geography**
>
> • understand, through the use of detailed place-based exemplars at a variety of scales, the key processes in:
>
> • physical geography relating to: geological timescales and plate tectonics; rocks, weathering and soils; weather and climate, including the change in climate from the Ice Age to the present; and glaciation, hydrology and coasts
>
> • human geography relating to: population and urbanisation; international development; economic activity in the primary, secondary, tertiary and quaternary sectors; and the use of natural resources
>
> • understand how human and physical processes interact to influence, and change landscapes, environments and the climate; and how human activity relies on effective functioning of natural systems

Fig. 1.1 Extract from the 2013 National Curriculum showing the slimmed down content requirements for geography (DFE 2013)

the nature and relevance of their subject and to ensure that there was a strong geographical element to their teaching. To convince teachers, the argument needs to be framed within a broad framework of ideas, linking the geographical knowledge taught in the classrooms with a broader set of aims of schooling. This is the challenge of this book and the notion of 'capabilities'. But this argument is not just restricted to geography in English schools.

The USA

The USA is not only one of the most economically developed countries in the world, but also has one of the highest amounts of money spent on each pupil per head of population (National Centre for Education Statistics 2017). Education policy that determines what goes on in schools is decided by each individual state, and sometimes even districts within states, resulting in a system in which school children do not receive an equal exposure to geography. As discussed in Solem et al. (2013), at a national level, there are 18 'national standards' of geography categorised into 6 essential elements: The world in Spatial Terms; Places and Regions; Physical Systems; Human Systems; Environment and Society; and the Uses of Geography.

However, these are voluntary, and each state creates their own policy on geography based on these guidelines. For younger students (Kindergarten to grade 5, ages 5–11), geography is integrated into topics and social studies. At middle school (grades 6–8, ages 11–14), 18 states require geography, either as a standalone course or as part of a combined geography and history course. Eleven states have no requirements for middle school geography; there are some individual districts within a further 22 states which may require geography. At high schools (grades 9–11, ages 14–18), 27 states require schools to teach geography or a geography/history course. Seven states have no high school geography requirements, whilst again individual districts in 17 states may require geography. Where geography is present, it is often part of a social studies programme or combined with history. The result of this is that many children in many states do not receive any geographical education at all.

Rohli and Binford (2016) discuss this in relation to the State of Louisiana. As they argue "as in most of the United States, geographic education in Louisiana remains in a precarious position due to economic constraints and a lack of appreciation in U.S. culture for the value of the geographic perspective" (p. 224). The implications of this across the country have been expressed by Martha Nussbaum (2006) in her work discussing the importance of education to create democratic citizens who have international perspectives on the world. As she argues:

> Americans are so resistant to serious learning about any other country. Because of America's size, wealth and power, they feel perfectly able to go through life without this learning. People in most other nations are less likely to sustain a comparable degree of ignorance. (p. 390)

This shows there is a need to articulate clearly the value of a geographical education to a wider audience of policy makers, teachers, parents and politicians, as this would then ensure adequate funding and training. A further challenge arises from the ways in which teachers in the USA are trained, even those where geography forms part of the curriculum. As Solem et al. (2013) explain:

> In the U.S., teacher preparation in many states gives only cursory attention to geography even though geography is present in state standards. This situation owes to the lack of geography courses offered on the campuses of many teacher education programs. Because of their inadequate preparation in geography, American teachers have long felt unprepared to teach the subject. (p. 217)

The challenge of geography education in the USA involves changing attitudes, funding regimes and teacher training to ensure all children receive a geography education. But the geography on offer to school children is varied and unevenly distributed. Work has already begun to address these challenges, building on the ideas in this book with the US 'National Centre for Research in Geography Education' launching its 'Powerful Geography' project (e.g. powerfulgeography.org). Dr Michael Solem is leading this project, and he was a partner in the GeoCapabilties 2 project (described in Chap. 5) and thus helped develop the conceptual understanding of GeoCapabilties.

Finland

Finland, in Northern Europe, is a country which often tops global polls for educational quality (e.g. PISA, The Programme for International School Assessment, as discussed in Tani 2014). The Finnish education system is covered by a National core Curriculum, which has a stated overall aim; "to secure the necessary knowledge and skills as well as encourage learning" (Finnish National Board of Education 2014b). The National Curriculum document (Finnish National Board of Education 2014a, b) covers aims, required content and assessment criteria at a variety of stages. Geography features as one of 20 subjects in the Finnish Curriculum, alongside social studies and environmental studies. Pupils in Finnish schools also develop a series of 'transversal competencies', which are developed alongside all their subjects. These include 'thinking and learning to learn', 'multiliteracy', 'ICT competence', and 'participation, involvement and building a sustainable future'.

During the first six years—in the primary school—geography is integrated in environmental studies together with biology, physics, chemistry and health education; at lower secondary level (grades 7–9), geography is taught as an independent subject. Geography is much more closely allied to the physical sciences in schools. Most geography teachers teach biology and geography; and the majority of them have biology as their major subject in their academic degree, as Kaivola and Rikkinen (2007) explain:

> [F]or most of the 20th century, geography... was linked with natural history (now called biology), these two subjects being the responsibility of the same specialist teacher. (p. 316)

Courses for upper secondary students include the Changing world, Blue Planet, Our Common World, Geomedia and local applied geography. As the curriculum document explains:

> Pupils study the Earth and its regions, nature, human activities, and different cultures. Aspects of natural, human, and social sciences are taken into account... to construct a cohesive overall picture of a diverse world and the way it works. Interaction between nature and human beings as well as its connection to the state of the environment are discussed... and a foundation is laid for understanding different regional views and conflicts on Earth. (Finnish National Board of Education 2014a, p. 2)

The Finnish geography curriculum is broad, contemporary and ambitious. Yet, despite the international reputation of Finland's curriculum, experts on the Finnish geography curriculum have recognised a series of challenges that seem to mirror those in England. One of these is the balance between knowledge and skills, in particular ICT (information and communication technology) skills; as Tani (2014) predicted before its implementation:

> [T]he (latest curriculum) shift will be from a knowledge-based curriculum towards a more skills-based curriculum. (p. 99)

A specific issue lies with the relationship between teaching geography and developing these broader curricular 'transversal competencies'. As Tani (2017) explains:

> The emphasis of phenomenon-based integration can be seen as a problem when complex (geographical) issues are taught. (p. 211)

With geomedia and Geographical Information Systems (GIS) mentioned specifically in the curriculum documents, and an increasingly digitised examination system, a danger is that geography becomes a vehicle for teaching ICT, rather than ICT being used constructively to enhance geographical learning. This is the opposite issue to that faced by the 'knowledge turn' of the English curriculum, but again highlights the need for clarity over the role of geographical subject knowledge in the school curriculum and the place of skills. Tani (2017) also identifies as a challenge changes to pedagogy which sees a promotion of 'learning to learn' as a core element of education. She perceives this move as "trendy" and a "threat" (Ibid., p. 211) to discipline-based teaching. The notion of 'learning to learn' seems to link to many of the aspirations of the English National Curriculum 'big picture' of 2009.

An increased clarity of the role of specialist knowledge, and the place of specialist teaching, has much to offer contemporary curricular debates in Finland. The ideas explored in this book around powerful knowledge, curriculum leadership of teachers and capabilities as a framework for thinking can be of use to provide a response to these debates; this work has begun in Tani (2017), who was a partner in the funded GeoCapabilities projects which helped develop the ideas (described in Chap. 5).

Setting the Scene: The Potential of a Geography Education

The geography on offer to school children throughout the world is varied, as illustrated by the examples from England, the USA and Finland. Yet young people growing up in the complex modern world need a rigorous geography education, and much work has already been done to express the potential of geography in schools. As Boulding (1985) argued:

> What formal education has to do is to produce people who are fit to be inhabitants of the planet. (Otherwise) young people are going to grow up and discover that we have taught them how to live in a world long gone. (Boulding 1985, p. 1)

With the contemporary challenges of climate change, poverty, globalisation and environmental degradation, geography is the subject which tackles these issues with students. Taught well by geography subject specialists, these contemporary issues could be taught in a way that will not persuade the students to arrive at a pre-determined view, but would enable the children to engage with a variety of knowledge and data to build up their understanding so they can come to their own view about these contemporary issues. Other subjects, including citizenship and PSHE, have started to contain the topics currently part of a modern geography education. Thus, there is still a need to ensure these contemporary challenges are seen as being "geographical", as

> [w]ithout a substantial geographical component, it is possible to argue that young people will be restricted in their capacity to make sense of the complex, unequal, fast changing and often dangerous world in which they live. (Lambert 2008, p. 207)

Geography is the subject that will enable children to develop the specific knowledge required to understand and make sense of these contemporary global issues.

The arguments and examples of school geography presented here set up a confusing picture, and, in 2009, Lambert (2009) described geography in education as being "lost in the post".[1] He meant the subject community seemed too concerned with arguments over definitions of content, knowledge, skills and assessment when really there was a much bigger argument to make about the aims, values and purposes of geography as a subject in schools. If this could be expressed, if the subject community was able to articulate why geography mattered in schools, then decisions about what to teach would be easier to make.

It was this grand thinking that led to the Geographical Association's (GA) (2009) manifesto for geography. The GA, the subject association for geography teachers, was moving discussions about the subject away from a content-driven set of facts to be seen more as a "curriculum resource" as they argue "contemporary challenges… cannot be understood without a geographical perspective" (GA 2009, p. 5). The manifesto was a "re-affirmation of geography's place in the curriculum" (Ibid., p. 5). It was an ambitious mission statement about the nature and role of geography in the school curriculum. It was aimed at inspiring teachers in schools where geography had been marginalised or lost to 'humanities', or where an outdated and boring set of 'factual delivery' of knowledge dominated classrooms. It was aimed too at teachers whose real-world case study examples of phenomena had not been updated. As Lambert (2009) in the manifesto urges "we may need to throw out crusty old favourites … in favour of… lessons that challenge students to make geographical sense of their own lives and experiences" (p. 1). This quotation was a response to those 'boring' and 'irrelevant' lessons (as identified by Ofsted 2008, 2011) which were dominated by dry facts, but it too can be misinterpreted as promoting a purely child-centred approach to education which downplays the importance of any sort of knowledge development. Lambert (2009) was able to move the debate in the geography education community beyond defining facts that needed to be learnt, but at this

---

[1] It was in Lambert's (2009) inaugural professorial lecture in 2009 entitled 'Geography Education: Lost in the post' that the idea of 'capabilities' was first alluded to.

time he had not grasped the significance of trying to define the educational aims of the subject and the importance of knowledge as part of these aims.

In an attempt to provide a set of international standards for geography education, the 'International Geographic Union' (IGU), a group of expert geography educationalists from a variety of nations across the world, created the 'Charter for Geographic Education'. Written in 1992, it is an attempt to provide a unified international voice for the potential of the subject and its teaching in schools. As they argue, they are:

- *Convinced* that geographical education is indispensable to the development of responsible and active citizens in the present and future world
- *Conscious* that geography can be an informing, enabling and stimulating subject at all levels in education, and contributes to a lifelong enjoyment and understanding of our world
- *Aware* that students require increasing international competence in order to ensure effective cooperation on a broad range of economic, political, cultural and environmental issues in a shrinking world
- *Concerned* that geographical education is neglected in some parts of the world, and lacks structure and coherence in others
- *Ready* to assist colleagues in counteracting geographical illiteracy in all countries of the world (IGU CGE 2016, p. 1)

Their document links geographical education to the Universal Declaration of Human rights, the work of the United Nations and gives advice on how geography can be approached in schools. The work of the IGU CGE (Commission on Geographical Education) is as relevant today as it was when the original document was written in 1992. The document attempts to do much of what this book also intends: to provide a robust defence of geography as a subject in schools and articulate a means to express it. The IGU CGE holds no political power, however, and remains as a set of guidelines. As we have seen from the experiences of children in the USA, England and Finland, exposure to geography in the classroom is varied and as such the aspirations of the IGU CGE have not been met in many nations.

The IGU CGE outlines how and why geography has the potential to enlighten and inspire all pupils but what has been lacking in the discourse is an overall conceptual framework to link geographical knowledge with a broader set of educational goals and values. This book explains how the 'capabilities approach' can be applied to curriculum thinking and could be a way to express this framework.

Setting the Scene: A Role for the Capabilities Approach

The 'capabilities approach' was created by the economist Amartya Sen (e.g. Sen 1980) and philosopher Martha Nussbaum (e.g. Nussbaum 2000). When applied to educational thinking, it attempts to focus on what human functioning and abilities can result from education, as an alternative to looking at what grades or scores a child can achieve. The capabilities approach to education seeks to understand the purposes of education from the perspective of the outcomes for a child. It asks what we want a young person to be able to 'be like', or to 'think like' or to 'do' as a result of their education. These qualities are the 'capabilities' of a young person. The thinking offered through an engagement with 'GeoCapability' attempts to express this through the subject of geography. As Lambert and Morgan (2010) argue "geography education can contribute to developing the capability of young people" (p. 63), suggesting the knowledge component of geography lessons has a significant role to play in helping young people to develop their capabilities.

Using the capabilities approach to think about the geography curriculum could provide an opportunity to articulate an underlying purpose to the subject; as Lambert and Morgan (2010) continue "capability provides a framework for clarifying… educational goals" (p. 64). An understanding of this can help teachers to devise engaging contemporary lessons with a strong geographical knowledge component, with a full appreciation of why they are doing it, and an understanding of how it leads to developing pupil's capabilities.

The notion of a capabilities perspective on geography education could have implications to how teachers and educationalists see themselves as geography educators, and to the role of geographical knowledge in education. A capabilities perspective on geography education could provide a means by which teachers can envision a contemporary curriculum and structure their professional thinking. It could provide a coherent argument for the inclusion of geography in a curriculum for decision makers in those countries where geography is under threat as a subject in schools. A capabilities perspective on geography education could invest school geography with its educational potential.

This book sets out to explore all these possibilities. 'GeoCapabilities' was developed and furthered through my own doctoral research, as well as two internationally funded research projects, both explained in later chapters. At the time of publication a third project is now underway. Educationalists from the American Association of Geographers and Helsinki University, Finland, were key partners in the project, as they felt a GeoCapabilities perspective could offer them some clarity and focus in developing geography education in their own countries. It is for this reason the experience of young people's geographical education in these countries has been outlined as part of this chapter.

Conclusions: Geography Education's Potential and the Capability Approach

This chapter sets the scene for the debates and arguments which follow. For many teachers and leaders already working in schools these debates will be familiar; for beginning teachers, they provide an introduction to some of the inherent challenges the profession faces. This book sets out the nature of these challenges but also offers means to think about curriculum in more positive ways that empower geography teachers to see the potential of geographical knowledge in terms beyond simply the passing of exams. The capabilities approach can provide a framework for this thinking. This book therefore has a series of specific aims:

– This book sets out a persuasive argument for a knowledge-led curriculum, built around knowledge-led capabilities within a subject-based curriculum.

- This book argues that the 'powerful knowledge' of geography as a subject in schools has a unique contribution to make to the education of young people beyond the passing of examinations and should therefore be a central tenet in the school curriculum.
- This book aims to encourage all geography teachers to re-think the role they play in the big picture of education, and thus show school leaders and policy makers the value of the subject and a subject-based curriculum in the face of competing curriculum pressures.

> **Questions to Consider:**
> 1. What is the purpose of schools to you? Have your ideas about the nature of schooling changed over time?
> 2. Think back to your own school days. What was the balance of knowledge and skills in the education you received? Do you notice a difference now?
> 3. How do you respond to the ideas of the Habits of Mind, the classical trivium and the National Curriculum?
> 4. Does it matter that 'geography' as a school subject is perceived differently throughout the world? Do these differences in the nature of the subject undermine attempts at developing an internationally coherent school subject?
> 5. How could you respond critically to the aims, aspirations and ideas of the IGU CGE charter for geography? Do they capture geography as an international discipline?

References

Association of School and College Leaders. (2016). *Survey Shows Damage of Teacher Shortages*. Retrieved March 17, 2018, from https://www.ascl.org.uk/news-and-views/news_news-detail.survey-shows-damage-of-teacher-shortages.html

Baker, D. (2014). *The Schooled Society: The Educational Transformation of Global Culture*. Stanford: University Press.

Bonnett, A. (2008). *What is Geography?* London: Sage.

Boulding, E. (1985). *A Guide to Curriculum Planning in Environmental Education* (D. Engelson, Ed.). Wisconsin: Department of Instruction.

Boyes, K., & Watts, G. (2009). *Developing Habits of Mind in Secondary Schools*. Heatherton, VIC: Hawker Brownlow.

Butt, G. (2008). Is the Future Secure for Geography Education. *Geography,* *93*(3), 158–165.
Carson, J. (2019, April 3). Fake News—What Exactly is It and How Can You Spot It? *The Daily Telegraph.* Online. Retrieved April 2019, from https://www.telegraph.co.uk/technology/0/fake-news-exactly-has-really-had-influence/
CIVITAS. (2007). *Corruption of the Curriculum—Press Release 11 June.* Retrieved December 2008, from http://www.civitas.org.uk/blog/2007/06/corruption_of_the_curriculum_p.html
Coughlan, S. (2018). England's Schools Face 'Severe' Teacher Shortage. *BBC News.* Online. Retrieved January 2019, from https://www.bbc.co.uk/news/education-45341734
Department for Education (DFE). (2013). *The Geography National Curriculum.* Retrieved December 2013, from https://www.gov.uk/government/publications/national-curriculum-in-england-geography-programmes-of-study
Department for Education (DFE). (2014). National Curriculum in England: Framework for Key Stages 1 to 4. Retrieved September 2018, from https://www.gov.uk/government/publications/national-curriculum-in-england-framework-for-key-stages-1-to-4/the-national-curriculum-in-england-framework-for-key-stages-1-to-4
Department for Education (DFE). (2016a). 'Specialist and Nonspecialist' Teaching in England: Extent and Impact on Pupil Outcomes. Retrieved March 17, 2018, from https://www.gov.uk/government/uploads/system/uploads/attachment_data/file/578350/SubjectSpecialism_Report.pdf
Department for Education (DFE). (2016b). *The English Baccalaureate.* Retrieved January 2016, from https://www.gov.uk/government/publications/english-baccalaureate-ebacc
Department for Education (DFE). (2017). Initial Teacher Training Census for the Academic Year 2017–18, England. Retrieved August 2018, from https://www.gov.uk/government/statistics/initial-teacher-training-trainee-number-census-2017-to-2018
Finnish National Board of Education. (2014a). National Core Curriculum for Basic Education 2014. Helsinki: Next Print Oy.
Finnish National Board of Education. (2014b). National Core Curriculum for Basic Education. Retrieved January 2019, from https://www.oph.fi/english/curricula_and_qualifications/basic_education/curricula_2014
Furedi, F. (2004). *Therapy Culture: Cultivating Vulnerability in an Uncertain Age.* London: Routledge.

Furedi, F. (2007). Introduction: Politics, Politics, Politics. In R. Whelan (Ed.), *The Corruption of the Curriculum*. London: CIVITAS.

Geographical Association. (2009). *A Different View, a Manifesto from the Geographical Association*. Sheffield: Geographical Association.

George, M. (2018). Reality of GCSE Reform is Pupils Studying Macbeth for Five Continuous Years. *Times Educational Supplement*. Online. Retrieved August 2018, from https://www.tes.com/news/watch-reality-gcse-reform-pupils-studying-macbeth-five-continuous-years

Habits of Mind Institute. (2018). Retrieved September 2018, from http://www.habitsofmindinstitute.org/

Hirsch, E. D. (1988). *Cultural Literacy: What Every American Needs to Know*. New York: Random House.

Holbrey, C. (2019). Geography Initial Teacher Education in 3 Days! The New PGCE 'PE with' EBacc Scheme. *Teaching Geography, 44*(2), 56–58.

Holt, J. (2015). 'Campaign for Learning' website. Retrieved March 2015, from http://www.campaign-for-learning.org.uk/cfl/LearningInSchools/L2L/index.asp

International Geographic Union Commission on Geographical Education. (2016). International Charter on Geographical Education. Retrieved November 2018, from http://geographiedidaktik.org/wp-content/uploads/2017/10/IGU_2016_def.pdf

JCQ, Joint Council for Qualifications. (2016). *Examination Results*. Retrieved January 2016, from http://www.jcq.org.uk/examination-results/A-Levels/2015

Kaivola, T., & Rikkinen, H. (2007). Four Decades of Change in Geographical Education in Finland. *International Research in Geographical and Environmental Education, 16*(4), 316–327.

Kreston UK. (2018). *Academies Benchmark Report 2018*. Retrieved March 17, 2018, from https://www.duntop.co.uk/pdf/Academies-Benchmarking-Report-2018.pdf

Lambert, D. (2005). Why Subjects Matter. *Nuffield Review of 14–19 Education and Training: Aims, Learning and Curriculum Series, Discussion Paper 2*. Presented 18 February 2005.

Lambert, D. (2008). Why are School Subjects Important? *Forum, 50*(2), 207–214.

Lambert, D. (2009). *Geography in Education: Lost in the Post?* (An Inaugural Professorial Lecture). London: Institute of Education Press.

Lambert, D. (2011a). Reframing School Geography: A Capability Approach. In Butt (Ed.), *Geography, Education and the Future*. London: Continuum.

Lambert, D. (2011b). Reviewing the Case for Geography, and the Knowledge Turn in the National Curriculum. *The Curriculum Journal, 22*(2), 243–264.
Lambert, D. (2016). Geography. In D. Wyse, L. Hayward, & J. Pandya (Eds.), *The Sage Handbook of Curriculum, Pedagogy and Assessment*. London: Sage Publications.
Lambert, D., & Morgan, J. (2010). *Teaching Geography 11–18 a Conceptual Approach*. Maidenhead: OUP.
Ledda, M. (2007). English as Dialect. In Whelan (Ed.), *The Corruption of the Curriculum*. London: CIVITAS.
Morgan, J. (2008). Curriculum Development in New Times. *Geography, 93*(1), 17–24.
Morgan, J., & Lambert, D. (2005). *Geography: Teaching School Subjects 11–19*. London: Routledge.
National Centre for Education Statistics. (2017). Retrieved December 2017, from https://nces.ed.gov/programs/coe/indicator_cmd.asp
Nussbaum, M. (2000). *Women and Human Development: The Capabilities Approach*. Cambridge: Cambridge University Press.
Nussbaum, M. (2006). Education and Democratic Citizenship: Capabilities and Quality Education. *Journal of Human Development, 7*(3), 385–395.
Ofsted. (2008). *Geography in Schools—Changing Practice*. Retrieved January 2009, from www.ofsted.gov.uk
Ofsted. (2011). *Geography: Learning to Make a World of Difference*. Retrieved December 2011, from www.ofsted.gov.uk
Qualifications and Curriculum Authority (QCA). (1998). *Education for Citizenship and the Teaching of Democracy in Schools*. London: QCA.
Qualifications and Curriculum Authority (QCA). (2008). *The Big Picture of the Curriculum*. Retrieved November 2008, from http://curriculum.qcda.gov.uk/key-stages-3-and-4/organising-your-curriculum/principles_of_curriculum_design/index.aspx?return=/News-and-updates-listing/News/Teaching-of-new-secondary-curriculum-begins.aspx
Roberts, M. (2010). Where's the Geography? Reflections on Being an External Examiner. *Teaching Geography, 35*(3), 112–113.
Robinson, M. (2013). *Trivium 21c: Preparing Young People for the Future with Lessons from the Past*. Bancyfelin, Carmarthen, UK: Independent Thinking Press.
Rohli, R. V., & Binford, P. E. (2016). Recent Trends in Geography Education in Louisiana. *Journal of Geography, 115*(5), 224–230.

Sen, A. (1980). *Equality of What? The Tanner Lecture on Human Values*. Delivered at Stanford University May 22, 1979. Retrieved November 2015, from http://tannerlectures.utah.edu/_documents/a-to-z/s/sen80.pdf

Solem, M., Lambert, D., & Tani, S. (2013). Geocapabilities: Toward An International Framework for Researching the Purposes and Values of Geography Education. *Review of International Geographical Education, 3*(3), 214–229. Retrieved December 2013, from http://www.rigeo.org/vol3no3/RIGEO-V3-N3-1.pdf

Standish, A. (2007). Geography Used to be About Maps. In R. Whelan (Ed.), *The Corruption of the Curriculum*. London: CIVITAS.

Standish, A. (2009). *Global Perspectives in the Geography Curriculum: Reviewing the Moral Case for Geography*. London: Routledge.

Standish, A., & Cuthbert, A. S. (Eds.). (2018). *What Should Schools Teach? Disciplines, Subjects and the Pursuit of Truth*. London: UCL IOE Press.

Staufenberg, J. (2018). PE Teachers Lured into Shortage Subjects via New Plan. Retrieved January 2019, from https://schoolsweek.co.uk/pe-teachers-lured-into-shortage-subjects-via-new-plan/

Tani, S. (2014). Geography in the Finnish School Curriculum: Part of the 'Success Story'? *International Research in Geographical and Environmental Education, 23*(1), 90–101.

Tani, S. (2017). Challenges for Geography Education: Digitalization, Integration and the Changing Role of the Teacher. *Terra, 129*(4), 211–222.

Tapsfield, A. (2016). Teacher Education and the Supply of Geography Teachers in England. *Geography, 101*(2), 105–109.

White, J. (2006). *The Aims of School Education*. Retrieved February 2011, from http://eprints.ioe.ac.uk/1767/

Young, M. (2003). Durkheim, Vygotsky and the Curriculum of the Future. *London Review of Education, 1*(2), 99–117.

Young, M. (2008). *Bringing Knowledge Back In: From Social Constructivism to Social Realism in the Sociology of Education*. Abingdon: Routledge.

2

Mapping a Curriculum 'Crisis'

Introduction

This chapter explores the first of the two major concepts that underpin the ideas in this book, the curriculum. The word 'curriculum' is readily used in schools. At its simplest level, it refers to the organisation of the school day and the way the timetable is envisaged. It is used to describe the 'academic' bit of school life, the subjects and knowledge the children learn. It is often accompanied by 'extra' (or 'co') curricular activities, which are those that focus on the broader holistic skills and competencies part of the school experience, such as the Duke of Edinburgh's Award scheme. The curriculum can be devised in a number of ways; subject based, skills based, competencies based and all of these are a means to structure what goes on in schools.

The word 'curriculum' is often used interchangeably with 'pedagogy' yet the two are distinct. Curriculum is the 'what' of teaching; it describes what we choose to teach children and the thought process that underlies why we teach it. Pedagogy refers to how the teaching is done, what actually goes on in the classroom to enact the curriculum and meet the various curricular aims. Yet the word 'curriculum' has a complex meaning.

The Concept of 'Curriculum'

The concept of 'curriculum' is one of the most significant and profoundly important ideas in educational discourse. As Lambert (2016) explains,

> the idea of 'curriculum' is arguably one of the very few powerful concepts genuinely to have emerged from the practice and study of education in modern times. (p. 395)

The term itself is derived from Latin for 'race course', as Lovat (1988) explains:

> [C]urriculum is, literally, a *course of action* designed to do a job, normally the job of educating an individual, a group or even an entire nation… A curriculum… is a *course to be run*. (p. 205 original emphasis)

This analogy suggests that a curriculum therefore requires a sequentially interrelated set of ideas that runs and develops over time. The origins of the word in contemporary education derive from Bobbit (1918), who argued that the very existence of a curriculum helps shed light on all interactions in the 'social engineering arena', how children gain and interact with knowledge and ideas. He was the first to explicitly link the 'ideal' of a school curriculum with a young person becoming an adult.

Stenhouse (1975) defined curriculum, arguing:

> As a minimum a curriculum should provide the basis for planning a course, studying it empirically and considering the grounds of its justification. (p. 5)

Thus, curriculum is a concept to be studied and understood.

Yet in schools, 'curriculum' has taken on a much more pragmatic definition, as an expression of how learning is organised. It defines the knowledge content as well as the skills, understanding, dispositions and ideals that students are to gain. For teachers, curriculum therefore has two meanings. One is the theoretical consideration of curriculum including its aims, why learning is designed in a particular way; what we want the children to be able to do by the end of their learning; and ideologies and the belief system which underlies the way teachers work. The second is the practical definition: how the school day is structured to enable children to learn. Teachers cannot do the second without a clear understanding of the first, and the relationship between the theory and practice is an important consideration.

Many attempts have been made to diagrammatically represent the various components of curriculum and in an attempt to show the relationship between the theoretical definition of curriculum and its more practical outcome, Lovat (1988) devised a model (Fig. 2.1).

Lovat's (1988) model suggests that curriculum theory and curriculum practice are closely interlinked. The theoretical base (in the centre of the diagram) which directly informs teacher practice is a product of three major considerations. First is the content base of curriculum, which includes subject knowledge and skills. This is the input to the curriculum. The second is the foundational base of educational disciplines, the understanding of educational discourse. This includes an understanding of how children learn, and the nature of assessments, progress and development of pupils. These combine to form the third consideration, the methodological base, how curriculum is designed and developed to enable children to engage with specific subject knowledge. This then leads on to the theoretical base, how the curriculum is conceptualised

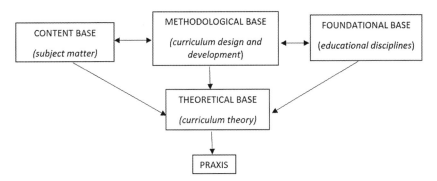

Fig. 2.1 A model of curriculum linking content, methodologies and education to inform curriculum theory, which directly informs practice. Redrawn from Lovat (1988, p. 212, www.tandfonline.com)

and what ideologies underlie their creation. These careful considerations then lead on to 'praxis', the practical aspect of the curriculum; the classroom teaching and learning. Teachers working in schools bring the curriculum to life with a clear and defensible understanding of how these curricular ideas were developed through a strong curricular conviction.

Yet Lovat (1988) also concedes that teachers do not have a clear understanding of 'curriculum'. Despite this long heritage of curriculum theory by educationalists, "we belong to a philosophical tradition which has tended to downplay the relation of theory to practice" (Lovat 1988, p. 206). Curriculum theory seems to have little impact on the actual school curriculum experienced by students since the 1980s. The reason, as Priestley (2011) explains,

> may be attributed to the tendency for curriculum policy to become more prescriptive since that time. Initiatives such as England's National Curriculum undermined teacher autonomy through prescription of content, and increasingly methods. (p. 225)

The National Curriculum, for decades, defined the knowledge content of subjects. The result has been a generation of teachers from the late 1980s through to the present day who have become more like 'technicians', delivering a pre-determined curriculum resulting from government policy (e.g. Biesta 2010) rather than reflexive professionals working within the deep grounding of their subjects and mediating more flexible, less prescriptive curriculum policies (e.g. Osborn et al. 1997). With the broadening out of the National Curriculum in 2007, subject knowledge became less prescriptive so teachers were in a position to develop their own curricula. This has enabled a broad range of content to infiltrate traditional subjects, and in some cases, the very existence of subjects has diminished as schools identify new ways to structure a curriculum to educate their young people.

In contemporary educational discourse, 'curriculum' has been conceptualised in many ways, by both Scott (2008), and Beck and Earl (2003). For Scott (2008), four key aspects of curriculum are important: objectives, content, methods and assessment. First are aims or objectives; this dimension considers the purpose of a curriculum and what it sets out to achieve either explicitly or implicitly. The content of subject matter focuses on knowledge, and what knowledge is selected to be taught and what is not. Methods or procedures is a consideration of pedagogy, or the way in which learning is organised, and this would include ideas such as discrete subject teaching or interdisciplinary lessons, types of learning activity, the use of textbooks and resource material. Evaluation or assessment is a way of checking if the aims have been met, either through 'summative' tests—such as national examination systems—or through more classroom-based 'formative' assessment (see the work of Dylan William for more on assessment for learning e.g. 2009).

These four dimensions are similar to the theories outlined by Beck and Earl (2003), who argue a curriculum can be defined in terms of three ideas: scope, content and structure. For them, the scope of a curriculum describes how schools transmit knowledge and has two aspects, the 'overt' scope and 'hidden' scope. 'Overt' refers to explicit curricular aims, clear objectives and organisational aspects of pedagogy such as how pupils are grouped for learning, and which different forms of learning are recognised. This is directly linked to Scott's (2008) idea of 'aims'. Beck and Earl (2003) also identify the 'hidden' curriculum, which relates to implicit aims of education, ideas about which values the school does and does not transmit. These values are not explicit, and are not made obvious to school pupils or parents, but are instilled into pupils through their time in school, and through the knowledge that they engage with. Gordon (1982) has expressed the hidden curriculum in terms of "attitudes, values, dispositions, (and) certain social skills" (p. 188) that underlie interactions between individuals within the school. The scope of the hidden curriculum is particularly wide, as Gordon (1982) reviews:

> [F]or example, we are told the hidden curriculum teaches kids to be passive (Silverman 1970); to be independent (Dreeben 1968); that conflict is unimportant (Apple 1971); that girls are inferior to boys (Fraziern and Sadkerm 1975); (and) that escalating consumption is a prime value (Illichi 1971). (p. 193)

Teachers, and those who create a curriculum, are in a position to use the 'hidden' curriculum to promote values and attitudes. More recently, the hidden curriculum has been discussed in relation to notions of 'politicalisation' (e.g. Marsden 1989) and 'corruption' (Whelan 2007, Standish 2007, 2009), with an increasing influence over the school curriculum being held by political groups and organisations.

Beck and Earl's (2003) thoughts about curricular content are similar to Scott's (2008) notions of content. For Beck and Earl (2003), "any curriculum is a selection from all the worthwhile knowledge which schooling could potentially transmit" (p. 14). Rather than discussing this from a knowledge perspective, and a discussion about what the content of a curriculum might be, Beck and Earl (2003) offer a series of principles which underlie the decision of what content to select. They argue that this should be based on

> children's and student's own interests and choices; economic relevance; vocational relevance; shaping national identity and allegiance; a humanistic conception... emphasising the value of knowledge and understanding for its own sake. (p. 15)

Beck and Earl (2003) here suggest that the curriculum should be determined by a set of principles and curriculum ideologies rather than simply a list of facts drawn up to be learnt. If the former is the basis on which to select curriculum content, those creating a curriculum, whether teachers in schools or governments or 'corrupting' influences, need to have an understanding of often conflicting ideological traditions that underlie the curriculum and this in part is why understanding curricular aims is a central concern in educational discourse.

Beck and Earl's (2003) final consideration of curriculum structure refers to how the content is organised:

> The nature of the elements that make up the curriculum and relationship to one another... for example, ... a set of discrete subjects, each separately timetabled and (often) taught by specialist teachers. (p. 15)

It combines Scott's (2008) third and fourth dimensions of curriculum as it combines both pedagogy and assessment. Scott (2008) seems to implicitly support a curricular structure of separate subject disciplines, taught in discrete timetabled blocks. Yet for Beck and Earl (2003), it does not necessarily follow that separate subjects form the basis of curriculum structure, and there is now a broad variety of curricular structures in English schools in which 'content' is organised and classified in different ways for students.

The aims of a curriculum refer to decisions taken at the outset of curricular design, and one way of expressing this is through a consideration of what we want young people to have achieved by the end of their education. Green (1980) differentiates between the secondary and primary 'goods' of education that occur as a result of education. Secondary goods, or 'outputs' are diplomas, certificates, awards and grades which can be measured and used to compare schools with other schools and educational institutions offering the same qualifications. The primary goods, or 'outcomes', are knowledge, skills and understanding developed by children whilst in schools. Outputs are measurable, whereas outcomes are intangible, more holistic understandings and skills. This distinction is significant, as discussions around 'GeoCapabilities' express the outcomes of an education, rather than outputs, and hark back to the educational ideals of Bobbit (1918) linking the curriculum to an educated adult.

Since the creation of the National Curriculum in 1988 due to the direct control of the school curriculum by successive British governments, school teachers and academics have not been able to engage with curriculum theory, and as such "the field of curriculum studies, at least in the UK, has declined in both status and practice in the universities and in the wider educational community" (Priestley 2011, p. 222). This lack of status for academic curriculum studies has also had an impact on schools, and Green (1980) argues that developed societies often put more value and emphasis on the outputs of education, rather than the outcomes of education, due to the ease of being able to quantify and measure examination grades and achievements. This has created a results-orientated school system in which the importance of exam achievement is a central focus of school curriculum and drives the ideology of schools, with the teachers mentioned in the previous chapter explaining this at a national education conference. League tables of schools based on examination results are published each year and these are hugely influential in ascertaining the success of a school. Yet examination scores only record what children have been able to achieve as part of the examination system and do not recognise any of the personal qualities that might derive from education, such as the development of values or morality. It also says nothing about the ability of that young person to think and work independently, to apply understanding to the real world, or to make decisions and choices for themselves about how to live. For Green (1980), the success of schools is based on a narrowly defined and restrictive set of easy to measure criteria rather than by any form of assessment of a broader range of personal qualities. This results focus has altered the entire discourse of curriculum thinking. As Carr (2004) explains,

> the most pressing problems facing educational practitioners were no longer the kinds of questions that, in its initial twentieth century embodiment, the philosophy of education had sought to address. Instead, they were narrow technical questions about how the externally imposed goals set for the educational system by the state were most effectively to be achieved. (p. 64)

This focus on outputs has resulted in teachers being encouraged to 'teach to the test', ensuring children have the facts and skills needed to pass nationally set examinations to retain the school's positioning in league tables.

Yet the 2007 curriculum rewrite and subsequent reforms loosened the government's control of the curriculum, allowing teachers to innovate in the classroom and devise their own curricula. Teachers had to re-engage with curriculum theory in order to devise meaningful lessons for their students. This could only have been achieved if teachers had a clear idea about the aims of a school curriculum, what it was trying to achieve and why; yet advice from central government on this was non-existent, and there was distinct disagreement among academics and professionals about what a curriculum in the first half of the twenty-first century should look like, and what it should try to achieve.

The Curriculum: From 'Problem' to 'Crisis'

The curriculum 'problem' was a term coined by Graves (1975), working in the field of geography education, when he identified the challenges in deciding what the aims and nature of a school curriculum should be. The

National Curriculum failed to provide adequate answers to the problem, and so the challenges were never truly resolved and as such led to talk of a curriculum 'crisis' at the start of the 2010s (Wheelahan 2010). The crisis took many forms: a crisis over the aims of what a curriculum should seek to achieve; a crisis in the balance of knowledge and vocational skills; and a crisis in the role of traditional academic subjects. The crisis has been picked up by a number of other writers. As Carr (2004) explains,

> amongst the questions that the contemporary educational discourse excludes are, of course, substantive philosophical questions about the fundamental aims and values that should provide the intellectual basis for contemporary educational policy and practice. (p. 57)

Part of the problem, as Carr (2004) identifies, is the discrepancy between the different groups of people, or stakeholders, who all have an interest in education, and who have different ideas about the aims of education, such as educational academics, politicians, teachers, parents and the public. As he continues:

> [O]n the one hand, we have a small academic community of educational philosophers whose members examine these issues in accordance with the canons of rational inquiry but whose arguments and conclusions have little practical effect. On the other hand, we have a diverse group of politicians, policy makers, teachers and other educational professionals who make and implement practically effective educational decisions but do so in a way which generally lacks intellectual rigour and in which serious and systematic reflection on the fundamental philosophical standpoint that informs their decisions is conspicuously absent. (p. 57)

This highlights the intellectual divide that exists between education academics, policy makers and teachers in schools. With the increasing prevalence of social media, ideas and opinions about education are created and shared widely, giving an audience to teachers who are able to discuss ideas readily without any research basis. A number of books have been published on education by teacher writers who often use their personal experience rather than any research basis to explain their ideas. What has been missing in these discussions is a shared ideology about the aims of a school education that are informed by research from the academic discipline of education, and enacted upon by politicians, teachers, school leaders, parents, pupils and society.

This ideological basis for a school curriculum could be expressed in the form of a school's 'ethos'. An ethos sets out what a school is trying to achieve; it is an embodiment of the school's aims. It is idealistic and could promote academic success, or develop the 'whole person' or a multitude of other broad aims of education. John and Osborn (1992) identify a clear link between a promoted school ethos and the attitudes and values of the pupils in terms of citizenship ideals, democratic rights and individual freedoms. Schools are often very keen to set out their ethos in promotional literature and websites, though Donnelly (2000) has argued that there can be discrepancy in what schools promote as their ethos and what is observed in the interactions of the school community. An ethos is the manifestation of an ideology, a belief about the aims of education, and different ideologies give rise to different curricular organisations. An ethos driven by academic success might create a curriculum that is different from one which promotes a more holistic skills-based ethos. In a piece of research into ideologies, Rawling (2000) summarises the main ideological traditions that have underpinned curriculum debates in the late twentieth century, and this is shown in Table 2.1.

The coexistence of these different ideological traditions underlies much of the current curriculum crisis. An education through a 'utilitarian' or 'vocational' ideology is an ideology which "establish(es) a clear link between... education and the needs of the economy" (Trowler 1998). A school curriculum with this ideology is aimed at skills training which is specific to the labour market of the time, and if certain skills were needed by sectors of the economy, teachers in schools would adapt their curricula to suit. The power and control of curriculum design is from industry leaders. Schools would take on a more vocational curriculum, offering trades and skills at the expense of traditional academic knowledge through subject disciplines.

In contrast, the reconstructionist or radical ideology is one "in which… (education) is seen as a vehicle for criticality and for transforming society" (Trowler 1998). Education from this perspective looks at children's actions and lifestyles and uses these as a means to envision other lifestyles to "facili-

Table 2.1 A range of different curriculum ideologies that influenced curriculum debates in the late twentieth century (Rawling 2000)

Ideological tradition	Characteristics
Utilitarian/informational	– Education primarily aimed at 'getting a job' and 'earning a living'. – A focus on useful information and basic skills.
Cultural restorationism (as promoted by the New Right in English policy making in the 1980s and 1990s)	– Restoring traditional areas of knowledge and skills (cultural heritage). – Providing students with a set package of knowledge and skills which will enable them to fit well-defined places in society and the workplace.
Liberal humanist (also called classical humanist)	– Worthwhile knowledge as a preparation for life; the passing on of a heritage from one generation to the next. – Emphasis on rigour, big ideas and theories and intellectual challenge.
Progressive educational (also called child-centred)	– Focusing on self-development or bringing to maturity the individual child/student. – Using academic subjects as the medium for developing skills, attitudes, values and learning styles which will then help them to become autonomous individuals.
Reconstructionist (also called radical)	– Education as an agent for changing society, so an emphasis on encouraging students to challenge existing knowledge and approaches. – Less interest in academic disciplines, more focus on issues and socially critical pedagogy.
Vocational or industrial trainer (Note: in some ways this cuts across all the other traditions)	– Provides students with knowledge and skills required for work. – Or use workplace and work-related issues as a stimulus for learning skills/abilities. – Or use work-related issues for questioning the status quo.

tate the construction of a new and more just society" (Schiro 2007, p. 6). Teachers' work through a 'radical' ideology would enable children to challenge rather than accept societal norms and encourage changes in behaviour and lifestyle. A reconstructionist agenda seeks to enable children to develop attitudes and views different from the socially accepted viewpoints, yet whose views are taught opens up further debates about who has power and control over the school curriculum. Encouraging young people to develop values and attitudes implicitly through the ideology of their education is linked to the ideas of the 'hidden curriculum'. Knowledge through traditional subjects plays less of an important role in this approach, with the aims of education being the taking on of radical beliefs. Fears of the radicalisation of young British Muslims through a radical curriculum became a key feature of the government's anti-terror drive in the 2010s (e.g. British Government 2018). This was a direct approach to prevent a radical ideological perspective being promoted in the school curriculum.

Ideologies can be expressed in the classroom through the ways teachers 'frame' knowledge for their pupils. The ideas of Basil Bernstein on curriculum 'framing' are particularly useful in this discussion. 'Framing' is a concept which identifies how knowledge is presented to children in the classroom; in the same way that an artist will carefully select the content of a picture to sit within a frame to tell a particular story, so a teacher can select appropriate knowledge and pedagogy for children to engage with in a classroom. As Bernstein (1971) explains,

> (the) frame refers to the degree of control teachers and pupils possess over the selection, organization, pacing and timing of the knowledge transmitted and received in the pedagogical relationship. (p. 88)

To frame the curriculum for children, into which ideological belief about education can be passed from teacher to pupil, teachers need to 'recontextualise' knowledge. The concept of 'recontextualisation' (Bernstein 2000) models the process through which teachers take academic knowledge from their specialist subject and translate it to enable children to access and engage with it. This 'translation' involves carefully selecting knowledge and then choosing the best pedagogy to enable children to engage with it to create meaning, yet it is during this recontextualisation process that ideologies can be expressed.

A further addition to the discussion of Bernstein's ideas about framing expresses the extent to which there is a prescription of content within a curriculum. In a curriculum that is 'strongly' framed, teachers have limited control over the curriculum. The content, pacing of lessons, type and timing of assessment are all pre-determined and teachers simply follow the curriculum plan. In a 'weakly' framed curriculum, teachers have more

2 Mapping a Curriculum 'Crisis'

autonomy over their practice; they are able to select content, go through this at their own chosen pace and devise suitable assessment, utilising the voice and opinions of students in the process. For well-trained subject specialists who are confident with the nature of their knowledge a weakly framed curriculum can provide an exciting opportunity for teachers to devise an enticing and relevant curriculum. Yet it is in a more weakly framed curriculum that external voices can enter the curricular frame.

The significance of how these ideologies can manifest themselves in the classroom has been part of recent research (Bustin 2018), in which groups of trainee geography teachers were asked to plan a lesson on the impacts of climate change through one of Rawling's ideologies. Despite the lesson being the same topic, the way knowledge was framed for the pupils, the nature of the lesson aims, activities and ultimately what the children learn were fundamentally different through each approach. Table 2.2 outlines three ideologies and the resulting lessons (from Bustin 2018).

The curriculum 'crisis' can be illustrated through the different ideologies that teachers have which can underlie the nature of the school curriculum, and how teachers interact with and present knowledge to students. Yet the balance between knowledge and skills in a school curriculum is another feature of the contemporary crisis.

Table 2.2 Curriculum ideologies and their impact in the classroom (Bustin 2018)

Cultural restorationism ideological perspective:
Lesson aims: To be able to describe and defend the UK government's position on climate change.
Lesson overview: Pupils carry out a 'treasure hunt' for UK policies on climate change. They then summarise their findings, summarising each of the main UK policies.
Pupils write and then give a political speech defending the UK's position on climate change.
Explanation and critique:
The trainees saw this perspective as being about asserting a 'UK-centric' focus on the geography curriculum. Thus in a previous era it might have been about defending the British Empire but here it is about defending a political position. Challenge and critique of the government do not to feature in this lesson, and this ultimately gives the pupils a very one-sided and uncritical view of climate change policies. The trainees would feel uncomfortable teaching this lesson as they planned it!

(continued)

Table 2.2 (continued)

Reconstructionist/radical ideological perspective: Lesson aims: To be able to know how we can all reduce the impacts of climate change. Lesson overview: Pupils are to work in small groups to create an advertising campaign about how they can reduce the impacts of climate change on a personal level and use this to create an 'action plan' to change their own and their families' lifestyles. Explanation and critique: The trainees saw this sort of lesson as trying to 'change the world', hence the idea of a poster and action plan. They felt this sort of activity might be alright at the end of a unit on climate change, but that it had to be done with an appreciation that reducing the impacts of climate change is not solely about lifestyle changes at the individual level. It requires a much more critical understanding about the roles of global trade and the impacts and policies of multinational businesses and governments. They felt the lesson was a bit geographically naive. **Child-centred ideological perspective:** Lesson aims: To be able to imagine what life must be like to live in an area affected by climate change. Lesson overview: Pupils would watch a short video clip or use an 'intriguing image' to show a child of a similar age to them living in a low-income country being affected by rising sea levels. Pupils have to create a role play interview with the child to identify how they might be feeling. Explanation and critique: The trainees used the importance of 'empathy' to devise this lesson, which is implicit in this approach. They felt this could be a great way to bring the realities of climate change to life in the classroom. What they were concerned about is the extent to which the interview could be realistic; they felt a child in a developing country would not have a sophisticated understanding of global climate change to be able to contribute knowledge to any interview. The trainees would want to ensure that the pupils used their geographical understanding to bring depth to the interviews rather than it being all about their acting abilities or their empathetic skills. In hindsight, they suggested it might be better to interview a politician from a country affected by climate change.

Mapping the Crisis: The Role of Knowledge in the Curriculum

A feature of the differing ideologies is the emphasis placed by each on knowledge. The acquisition of academic knowledge is a key feature of any educational system (e.g. Young 2008); even an education system based around developing children's skills and values still requires knowledge as a means to inform debate. Yet in contemporary education, there is an antipathy between the knowledge dominant ideologies of the cultural restorationists and liberal humanists, and the child-centred ideologies of the progressive educationalists (see Table 2.1), in which knowledge in the curriculum is deemed less important.

Much of the current knowledge crisis relates to the way knowledge is produced, with an antithesis between 'objectivist' or scientific knowledge and the role that people play in socially 'constructing' knowledge (e.g. Young 2008). Drawing from epistemology, the idea of objectivism suggests that there is a set of knowledge that is universally verifiable and testable. The role people play in this knowledge is 'discovering' it through reliable and replicable methodologies. As Trigg (1973) distinguishes, "a fundamental distinction must be drawn between the way the world is and what we say about it, even if we all happen to agree. We could all be wrong" (p. 1). For Trigg (1973) objectivist knowledge exists and describes the way the world 'is' and Dawson (1981) agrees, arguing knowledge should not rely on opinions of people despite them being labelled as 'experts'. As he argues, "It is difficult to disagree with the objectivist contention that there is more to truth than the opinion of the majority or of a hegemonic social group" (p. 415). Yet there is a distinction between 'objectivist' knowledge that has been discovered by communities of specialists, who work with specific methodologies, and knowledge that is presented as 'truth' without any disciplinary basis. It is this latter type of knowledge that Young and Muller (2010) dismiss as being "under socialised" (p. 14), as people have not been involved in creating or challenging this type of knowledge.

An alternative stance on knowledge production is 'social constructivism', which argues all knowledge is claims made by people. These various 'voice discourses' (Moore and Muller 1999) include those from scientists working to a strict methodology to folk traditions and personal experiences. Postmodern approaches to the sociology of knowledge give these types equal status (e.g. Usher and Edwards 1994); under this view, there is no such thing as 'better' knowledge, just alternatives. But Young (2008) asserts "there is … only the power of some groups to assert (that) their experiences should count as knowledge" (p. 5). For Young (2008), legitimate knowledge can be produced by societal elites, leading to the creation of what he once dubbed "knowledge of the powerful" (Young 1971), where powerful people and groups create what becomes accepted as knowledge in the scientific communities and popular culture. In later work, Young and Muller (2010) argued this type of knowledge production is "over-socialised" (p. 14) as it relies on the attitudes of people, their opinions and ideas, more than deriving from a testable, set methodology.

As an alternative to 'under socialised' knowledge production, and the 'over socialised' constructivist knowledge, Young and Muller (2010) argue for a 'social realist' approach. In social realism, knowledge is 'realist' in the sense that it recognises that real knowledge exists, whether it has been 'discovered' or not. The 'social' element of social realism identifies that all human knowledge is in some way socially constructed, including that which is from expert communities. A further distinction of social realism is made with regard to Maton's (2010) discussion of 'knower structures'; that for every knowledge structure, there is also a knower structure (p. 161). Knowledge becomes specialised not only in terms of what is known but also by who is knowing it and how they are knowing it. 'Social realism' argues that some knowledge is 'better' and more reliable than others. Rather than the societal elites creating this knowledge, it is generated through 'epistemic communities' of subject experts who create knowledge according to the methodologies of that community of experts. According to this approach,

> social realism understands knowledge as emergent from the specialised collective practices of knowledge generation within epistemic communities. (Firth 2011, p. 293)

This knowledge therefore "relies on a regulatory rather than an absolute notion of truth" (Ibid., p. 293); the means for creating new knowledge is through following a regulatory pattern of values and norms developed within knowledge disciplines, rather than there being some form of objectivist, absolute truth. As Young (2008) argues, "there are rules, codes and values associated with different specialist traditions which make well-grounded claims about knowledge and how it is generated and acquired" (p. 63). It is the specialist nature of the knowledge claims that gives it its status and validity as social realist knowledge. New scientific knowledge can only be created by groups of scientists following scientific methodologies; historical knowledge can only be generated by historians following the norms that have developed in that discipline.

The ways knowledge is produced has implications for its role in the school curriculum, and how choices are made about what to teach young people. It is through these differences that the contemporary 'knowledge crisis' of the curriculum can be best expressed. Young and Muller (2010) identified a framework to express these differences, which can also express the challenges of differing curriculum ideologies outlined earlier, although their work was not in response to this. Their framework identifies three possible curriculum 'futures': 'Future 1' (F1), 'Future 2' (F2) and 'Future 3' (F3). Each alternate future seeks to identify what a school curriculum would be like with a different emphasis on the importance of knowledge or skills. These 'futures' express the nature of the knowledge and skill input to a curriculum, and they are not concerned at all with pedagogy or how teachers teach. Some of the curricular futures do lend themselves to particular pedagogies but their work is concerned with what is being included in a curriculum, and not how that is being taught. Rather than existing far into the future, many schools in the 2010s, and in the past, have exhibited curricula that can fit into one of the tripartite descriptors.

Mapping the Crisis: A Knowledge-Led 'Future 1' Curriculum

The cultural restorationist and liberal humanist ideologies (Rawling 2000, Table 2.1) are both similar in the importance they place on learning knowledge. Schiro (2007) argues proponents of these ideologies

> believe that over the centuries our culture has accumulated important knowledge that has been organised into academic disciplines… The purpose of education is to help children learn the accumulated knowledge of our culture. (p. 4)

Through this ideology, subject knowledge is the central consideration of school teachers, and discussions about what is being learnt are more significant in informing practice than how learning takes place. This belief is the central part of an F1 curriculum.

In an F1 curriculum, knowledge is still created by 'epistemic communities' of experts, but unlike a social realist approach once this knowledge has been created it is "treated as largely given, and established by tradition" (Young and Lambert 2014, p. 59). In F1 schools knowledge is uncontested. Knowledge is regarded as something to be learnt and repeated, and to be transmitted to those capable of achieving, rather than questioned by learners and engaged with. An F1 curriculum appears to be related to the ideas of knowledge as described by Hirsch (1988), when he describes "what every American needs to know", a list of facts and concepts and "background knowledge (for) necessary functional literacy and effective national communication" (pxi). As children get older, the amount of knowledge they learn in schools increases, and this provides the basis and structure of the school curriculum.

What an F1 curriculum also promotes is the idea of the existence of academic subjects in schools. 'Core' knowledge (a term used in the National Curriculum debates of 2007) is developed by those working in academic disciplines over time and as such the discipline becomes the structure through which young people engage with knowledge. Teachers in schools are subject specialists, graduates of the subject discipline they teach, and they deliver, uncritically, this 'core' knowledge to children. Children learn this knowledge, being tested on what they can remember at various stages. An F1 curriculum is described in Young and Lambert (2014) as the experience of British schools up to the 1970s, and to the current curriculum of some grammar or independent schools today. Some discussions in 2010, about the return of 'core knowledge' into secondary school National Curriculum, could be seen as a return to an F1 curriculum, with the Education Secretary of the time Michael Gove and his minister Nick Gibb seemingly influenced by the ideas of Hirsch (1988) and his contemporaries. These ideals seem to have continued into discussions around the GCSE (General Certificate of Secondary Education) and A Level reforms of 2016–2017.

The existence of an F1 curriculum leads to a question of who is in control of knowledge selection, similar to a discussion in Beck and Earl (2003). Not every piece of knowledge can be taught to pupils in schools, as this would be an impossible task, not least as new knowledge is constantly being created, and so a selection needs to be made from all this knowledge about what should and what should not be taught. This decision relates to who has the power to make these choices, and the ideology to which they subscribe. In an F1 curriculum, there is a set 'core' of knowledge that is undisputed, and it is this knowledge which forms the curriculum content. This is the 'knowledge of the powerful' that underpinned much of Young's (2008) critique of socially constructed knowledge. It was the 'social elites' who decided upon the content of the curriculum and it is this which forms the 'canon' of academic content with which children need to engage. Young (2008) argued that many children were not given access to this knowledge, particularly when referring to the school system that separated the brightest children for grammar schools and the rest into secondary modern schools, before the introduction of the comprehensive school system.[1] Understanding who does have power and control over curriculum content is still a contentious issue in educational discourse. The government, through the National Curriculum, controls the curriculum up until the end of key stage 3, but awarding authorities, leading textbook series and resources for teachers as well as teachers themselves also all have attitudes, values and beliefs about what a contemporary curriculum should be like and these factors all influence the nature and type of knowledge found in school lessons.

The corollary to an F1 curriculum is one which downplays the importance of knowledge and which leaves curriculum time open to pursue activities which are designed to help children to develop skills and competencies. This position is similar to Young and Muller's (2010) F2 curriculum, which expressed the role of skills in a child-centred, 'aims'-based curriculum.

---

[1] The 'comprehensive' system was introduced in 1965, but many schools remained as grammar schools with different areas operating different systems. In the 2010s, with government support, some existing grammar schools expanded to new sites and welcomed more children.

Mapping the Crisis: An Aims-Led 'Future 2' Curriculum

In Young and Muller's (2010) F2 curriculum, the role of knowledge is weakened at the expense of skills and vocational considerations. An F2 curriculum may be allied to Rawling's (2000) child-centred ideology (see Table 2.1), in which the perceived needs of the children become the central focus of the curriculum. In F2, "curriculum boundaries between subjects (are) weakened, as new forms of interdisciplinary studies (are) introduced…the curriculum (becomes) open to leisure, sports and other community interests" (Young and Lambert 2014, p. 60). As Pring (2005) identifies,

> subjects merge and alter….and there are new subjects like media studies, information technology, business and leisure studies and sports studies. Road safety and motor car maintenance are now becoming subjects. (p. 1)

Some voices are even calling for a school system to be devoid of any subjects, as Pring (2005) continues:

> [T]here seems to be a deep chasm between those who see the curriculum to be essentially constituted of subjects and those who want very different principles of organisation (practical activities, themes, interests, personal agendas, etc). (p. 2)

These new curriculum principles of organisation are the essence of an F2 curriculum. Pedagogic considerations overtake discussions about knowledge, and a child-centred education results, with a focus on vocational training and skills. Knowledge becomes material with which to engage with other ideas. Subject disciplines become arbitrary, or even a distraction, as subject teachers are asked how their subject contributes to F2 concerns such as promoting healthy living, developing particular values or offering careers advice. These ideas are spliced into a curriculum at the expense of subject knowledge. Young and Lambert (2014) explain how an F2 curriculum developed in British schools "as part of policies of social inclusion and widening participation" (p. 60) with the 'comprehensivisation' of the British school system in 1965. With a broader ability range of children in schools, it was deemed that the sort of 'knowledge of the powerful' (Young 1971) found in the grammar schools was not appropriate for all, and alienating to many, and so a more vocational education awaited students in the new comprehensive schools.

Yet this is not just a historical debate. Contemporary writers (e.g. Reiss and White 2013) propose a maintenance of what they call an "aims based curriculum", which bears the hallmarks of an F2 curriculum, whereby they argue that the purpose of schools is to "equip each child to lead a life that is personally flourishing and to help others to do so too" (p. 1). They do not promote subject-specific education but start with social, overarching aims of education, then use this to create a curriculum. As Beck and Earl (2003) explain,

> children and young people should be offered a curriculum which … includes… humanistic, aesthetic, social, political and moral education as vitally important elements (not tokenistic add ons). (p. 20)

These elements of students' personal growth and development are most important through this ideology. Pedagogical considerations would therefore be at the forefront of lesson design, with 'thinking skills' activities dominating teacher activity; the thinking skill (such as 'asking questions') becomes the aim of the lesson rather than a tool to develop subject knowledge. It also means that teachers become preoccupied with ensuring lessons are accessible, differentiated to varying learner needs, with a broad range of learning activities at the expense of discussions about what is being taught and why. It removes the importance of subject knowledge from teachers' curricular discussions. The F2 curriculum is exemplified by the 'learning power' philosophy outlined in Chap. 1, which leads to the 'learnification' of the curriculum (as Biesta 2012 would describe). Through this curriculum vision, a 'love of learning' becomes the end point of curriculum in itself, rather than being the means to develop knowledge.

Another overarching curriculum structure is the Habits of Mind work (e.g. Boyes and Watts 2009), introduced in Chap. 1. This is a manifestation of F2 thinking across a school. Whilst these Habits have noble intentions, how they are integrated into a school is critical in ensuring their success. A direct interpretation, where the development of these Habits becomes the central aim of lessons, could be indicative of F2 curriculum thinking. Habits are developed at the expense of subject knowledge, and thus subjects and the Habits become conflicting means to organise a curriculum. This leads to teacher confusion. If the Habits are integrated as a means for pupil reflection, rather than for teacher aims, then they can coexist with a subject-based curriculum. Thus, it is how Habits of Mind can be integrated within a subject-based curriculum that is key to its success in finding a unifying concept for the curriculum.

Another consideration of an F2 curriculum is a focus on the role schools play in developing moral citizens (e.g. the work of Wilson 1990). As Gutmann and Thompson (1996) argue:

> Schools should aim to develop their students' capacities to understand different perspectives …and engage in the give-and-take of moral argument with a view to making mutually acceptable decisions. (p. 395)

This approach to moral education focuses on taking children beyond the learning of facts, but being active and responsive to the facts they learn. As Wilson et al. (1967) continue,

> the real point of a moral argument is not to examine facts and logic: it is to be able to react psychologically to the facts in a more efficient or discriminating or honest way. (p. 65)

According to this view, knowledge of the issue is unimportant; it is the reaction to it that is key. Developing a sense of morality in young people could be an explicit aim of schooling, or could be part of the 'hidden curriculum', to encourage children to think and respond in prescribed ways. Yet the role that teachers play in moral education has been questioned by Haydon (2000), who argues:

> I think it is fair to say, however, that many teachers are not sure what role, if any, they have in moral education, and may tend to avoid the terminology of 'moral education' and 'morality' in their own discourse. (p. 356)

How teachers approach the development of moral education is therefore the subject of much debate. Wilson et al. (1967) advise:

[T]he pupil should not merely be presented with a series of alternative moral views and allowed to choose between them. This could amount to a form of 'window shopping' with no criteria for reasonable choice being given (p. 253)… (instead) educating a person in any 'form of thought' or 'department of life' involves… encouraging pupils not merely to believe certain right answers but to make up their own minds. (p. 251)

Encouraging independent thinking to engage with and develop morality is often at odds with the needs of a prescriptive, examinable content output of schools.

In practice, teaching moral issues without indoctrinating students into a set of values can be a challenging task, as teachers are also moral beings, with their own set of religious and moral ideas which can often be passed on to the students. The words that teachers use in the classroom and the way they react and respond to ideas can often convey beliefs and values that are passed, often unknowingly onto students, particularly during the recontextualisation of subject knowledge, where values can enter the curriculum. Teachers have to strike a balance between introducing pupils to a defensible framework of moral beliefs and practices, and indoctrinating children to believe a particular set of values based on their own ideologies, particularly if they subscribe to a 'radical' curriculum ideology (see Table 2.1). As Harrison (1977) argues "teachers do not teach physics, mathematics or history in order to convey their own scientific or political biases, since this way lies indoctrination" (p. 56). Teachers who do 'indoctrinate' their pupils by encouraging them to take on viewpoints can be accused of being "morally careless" (e.g. Morgan and Lambert 2005) in the classroom. The extent to which it is possible to avoid being morally careless is problematic, as McLaughlin (2003) argues "education cannot be value free" (p. 137). Teachers have the challenge of enabling children to engage with morals and values whilst at the same time not indoctrinating them into one narrow, specific view point.

A further challenge of moral education is found in the debate about whose morals and which values are to be taught and promoted, and which are not. As Harrison (1977) asks, "where do they (the morals that are taught) come from, what is their authorisation, how are they defended as the logical and right choice?" (p. 59). These fundamental questions do not have set answers, though in an attempt to provide an answer, Tubb (2003) distinguishes between 'public' and 'non-public' values. Public values are those values and morals which are deemed universal, which the vast majority of the population is in agreement over. They are non-arguable and would include ideas such as 'murder is wrong', and 'kindness toward other people is a virtue'. These values are ones that all teachers would promote in schools, often explicitly in the form of set aims, but also implicitly in the way teachers work and behave in schools. 'Non-public' values are those contentious issues that require discussion and thought. They may result in students developing a range of ideas and opinions. Examples of non-public values include the differing perspectives on the ethics of foxhunting or abortion. The problem is that there is no clear-cut definition of what constitutes public and non-public values in schools. What one teacher may consider a public value another may consider being debatable. Who is able to make this choice is also a key consideration for schools; it could be a classroom teacher's responsibility or a value promoted by a whole school as part of its ethos. In a National Curriculum, the role of the government is also significant in deciding which values should be promoted in schools. This makes moral education a challenging task for teachers.

It is the promotion of values in schools under the false auspices of a moral education that has led some writers (e.g. Whelan 2007) to talk of overt politicisation of schools, or 'corruption', the promotion of party political views within the classrooms in schools. For school geography, Standish (2007) argues that 'geography used to be about maps' but has now become a vehicle for pro-environmentalist, anti-capitalist sentiments (see my take on his discussions in Chap. 3). Furedi (2007) was even clearer, arguing:

> [O]ver the last two decades the school curriculum has become estranged from the challenge of educating children (p. 1)... increasingly the curriculum is regarded as a vehicle for promoting political objectives and for changing the values, attitudes and sensibilities of children. (p. 3)

The politicalisation of pupils is a potential consequence of a morally careless, F2 curriculum.

The knowledge-led F1 school is one in which knowledge is organised into discrete subject disciplines and 'delivered' uncritically to students in classrooms. The child-led F2 school puts greater emphasis on the perceived needs of the children and the pedagogical considerations required to develop moral values and vocational skills.

The debate so far has suggested that these two alternate futures are distinctive, and in a sense this is why the curriculum can be said to be in 'crisis'. Yet Young and Muller (2010) have envisioned an alternative, an 'F3' curriculum, and this view might help move discussions beyond the talk of 'crisis' in the contemporary curriculum. GeoCapabilities, the focus of this book, is a possible means to express an F3 curriculum within the subject of geography.

Beyond the Crisis: 'Powerful Knowledge' and a Future 3 Curriculum

In an attempt to move discussion beyond the distinctive and irreconcilable ideas of an F1 and F2 curriculum, Young and Muller (2010) and Young and Lambert (2014) have identified a 'Future 3' (F3) curriculum. An F3 curriculum is knowledge led, and which still respects the importance of subjects, but unlike F1 this knowledge is not static and cannot simply be read in a book or found out online. It also respects the need to develop young people's skills, including a set of values and morals, but in a way that is based on understanding, and therefore relies on knowledge. An F3 curriculum, as Young and Lambert (2014) argue,

> points to a new and always changing balance between the *stability of subject concepts* (implicit and over emphasised in F1 and underemphasised in F2), *changes in content* (underemphasised in F1) as new knowledge is produced and the *activities involved in learning* (overemphasised in F2). (p. 68, original emphasis)

An F3 curriculum therefore attempts to provide an alternative to those advocating an F1 or an F2 curriculum.

The type of knowledge that is implicit in an F3 curriculum is 'social realist' created through epistemic communities of subject experts, which is different from the inert and 'given' nature of an F1 curriculum. It is more dynamic, and the curriculum is as much about introducing children to the epistemic rules of the subject discipline enabling 'epistemic access' to the knowledge, as it is about learning specifics. As Young (2008) explains, F3 creates

> a curriculum space where learning is as much about learning to navigate and negotiate knowledge, its communities, practices, relationships and its ways of constructing objects/subjects as it is about learning particular subject concepts and processes. (p. 308)

In schools, this would mean studying geography is about learning to think like a geographer, understand how geographical knowledge is created, debated and argued over and not simply about learning geographical facts. The same is true for other subjects, learning to think like a mathematician, historian or linguist rather than simply learning a set of facts associated with the subject.

Powerful Knowledge

The type of knowledge that enables this epistemic access to subjects is what Young (2008) terms 'powerful knowledge', and it forms the basis of an F3 curriculum. In F3, all knowledge in schools is socially constructed by people and communities working within the distinct boundaries offered by subject disciplines; knowledge is open to debate, challenge and discussion by subject experts, hence its status as socially realist knowledge. Powerful knowledge has therefore also been called 'powerful disciplinary knowledge' (e.g. Lambert et al. 2015) to highlight the important role that subject disciplines play in creating this type of knowledge. Powerful knowledge is not simple, is not 'everyday' knowledge but requires deep thought and consideration, and this is what makes it distinct from the sort of knowledge envisioned in an F1 curriculum or from the list of facts identified by Hirsch (1988). Powerful knowledge is complex, often

Table 2.3 The key features of 'powerful knowledge', based on Young (2008)

Powerful knowledge is:
- Created, argued over and considered within academic disciplines according to the norms and values of that discipline, thus it is 'specialised' knowledge.
- It represents the 'best' knowledge available in that subject, created and argued over and as such it is evidence based.
- It is not given; it can be usurped by 'better' knowledge, can be open to constant reworking and debate by disciplinary specialists.
- It is not 'everyday' knowledge but requires deep thought and sustained engagement.
- The development of powerful knowledge from a subject specialist teacher provides a rationale for a subject-based curriculum.

abstract and requires sustained engagement to enable understanding. Table 2.3 outlines the key features of powerful knowledge.

Powerful knowledge is knowledge that children cannot access at home and which they have to attend school to engage with. As Young and Lambert (2014) assert, powerful knowledge

> is distinct from the common sense knowledge we acquire through our everyday experience… it is systematic, its concepts are systematically related to each other in groups that we refer to as subjects…and it is specialised. (p. 75)

For Young (2008), access to powerful knowledge is the reason why children go to school. The acquisition of powerful knowledge could be seen, therefore, as an aim of education in its own right. Subject specialist, qualified, professional teachers are key to the process. Teachers enable children to access powerful knowledge. Teachers are trained in subject disciplines, are experts in their subject, and it is this which makes the knowledge they have 'powerful'.

Powerful knowledge links to the sort of knowledge promoted through the 'liberal humanist' ideological perspective (Rawling 2000, Table 2.1), which has an emphasis on rigour, big ideas and the best of what has been thought in a subject. Powerful knowledge takes this thinking further by identifying that knowledge is never static, but always open to challenge and change.

For Young (2008), talk of powerful knowledge marks a re-statement of his relationship with knowledge, as explained in Morgan (2015). In his earlier writing of 'knowledge of the powerful' Young (1971) described knowledge as being the preserve of the elite, decided by and passed on from those in socially privileged positions to the next generation of powerful people. It was this passing on of knowledge that retained societal inequality through the grammar and independent school systems, with those children who ended up at secondary modern schools who were therefore denied access to this knowledge. The implication of Young's (1971) early work was to devalue knowledge as simply being part of an outdated elite society. His ideas were hugely influential in the sociology of education and perhaps in part a reason for the rise of an F2 curriculum in many schools. The implication of Young's (1971) work was that access to traditional subject knowledge was unequal, sometimes alienating and therefore unfair. One response was to offer a more child-centred, more accessible curriculum that all children could engage with. This resulted in a move in schools from an F1, elitist 'knowledge of the powerful' education system to an F2 curriculum with an emphasis on skills, but a lack of knowledge. When assessing the impact of his 1971 work, Young (2008) realised that what was needed to reduce inequality was not the complete removal of knowledge from all school curricula, but a way for all children to access what had in the past been seen as the preserve of the elite. According to Young (2008), all children, irrespective of the school they went to and irrespective of their socio-economic background, needed access to knowledge. Hence, Young's position was re-stated from writing about 'knowledge of the powerful' (in 1971) to that of 'powerful knowledge' (2008). Powerful knowledge therefore became part of his 'F3' curriculum vision.

Young's (2008) work is about curriculum and not pedagogy; he was interested in what goes into the curriculum and not how it is taught. Yet for Roberts (2014), powerful knowledge can only ever be powerful if it is taught well. As she argues:

> Whatever knowledge is selected and justified, it is only potentially powerful. Students do not necessarily learn what they are taught; they do not simply acquire knowledge because it has been prescribed in a curriculum. School knowledge remains inert if students are not motivated to learn it and if they cannot make sense of it in some way for themselves. (p. 204)

Table 2.4 Key features of powerful pedagogy (Roberts 2013b)

Powerful pedagogy:
- Critical pedagogy
- Recognise political nature of issues
- Ask questions that challenge status quo
- Probe ethical issues
- Expose hidden meanings of data
- Consider underlying political and economic structures

This is where there is a need for what she calls powerful pedagogies. A powerful pedagogy can enable pupils to not only understand the knowledge that is being taught, but to understand the political and ethical issues inherent in the knowledge itself. It may not be explicit, but the way a teacher thinks about the knowledge and the language they use in the classroom can help pupils to develop powerful disciplinary knowledge. Table 2.4 outlines the key features of powerful pedagogy (from Roberts 2013b).

Margaret Roberts' work in geography education (e.g. 2003, 2013a) has focussed on the 'enquiry approach' as a means to help pupils engage with geographical knowledge in meaningful ways. Through this approach children engage with questions which teachers help them to answer. As she says:

> In school geography, the use of enquiry-based approaches to learning can give students access to powerful ways of geographical thinking, by helping students understand the nature of geographical knowledge. (p. 204)

It is through an enquiry approach that children can access powerful knowledge and it is this approach that could be considered a means to develop powerful pedagogies in the classroom and a means to envision an F3 curriculum.

Conclusions

This chapter explored why the concept of curriculum is such a contested notion in education and therefore said to be 'in crisis'. A range of curriculum ideologies can help position people's relationships towards education and its aims, but the main arguments in the debate presented here have been around the role of knowledge in the school curriculum. The arguments have been framed around two alternatives; a 'knowledge heavy' F1 curriculum in which facts are passed down uncritically from teacher to pupil and a 'child-centred' F2 curriculum which focuses on generic skills and children's development at the expense of knowledge. Neither of these curricular visions is particularly appealing to educationalists but they do represent two extremes of the debate. To move beyond this, the notion of a 'powerful knowledge' led F3 curriculum produces a curricular vision in which knowledge plays a central role in curriculum thinking, but it is more ambitious than the sort of uncritical knowledge present in F1 thinking. It is powerful knowledge, created by subject experts through 'disciplined' thought processes, taught through an engagement with powerful pedagogies. It is this that is at the heart of an F3 curriculum.

Questions to Consider:

1. What does 'curriculum' mean to you?
2. Think back to professional development/INSET (In Service Training) courses you have attended or given to staff focussing on classroom activities. How many of these have been about pedagogical issues (how to teach) and how many about curriculum issues (what to teach and why)? Is there an imbalance and if so does this matter?
3. Can you identify your own ideological approach to education, based on Table 2.1? Think of a lesson/topic you currently teach (climate change/ Fairtrade/Global development). Re-plan the lesson(s) from a different perspective. How does this change your approach to that topic? Why might understanding of these ideologies matter?
4. How would you define the 'ethos' of your school and how does the curriculum support it? Does any of it form part of a 'hidden' curriculum?
5. Can you identify what the 'outputs' and the 'outcomes' of your school curriculum are? Which are more significant (outputs or outcomes) to you; to your senior or middle leadership colleagues; to parents; to Governors and to pupils? If the significance is different, what implications might this difference create?
6. What is the balance between knowledge and skills in your school curriculum? Can you identify with any of the approaches as expressed through the Future 1 and 2 heuristic?
7. What does 'powerful knowledge' and 'powerful pedagogy' mean to you?

References

Apple, M. W. (1971). The Hidden Curriculum and the Nature of Conflict. *Interchange, 2*, 27–40.
Beck, J., & Earl, M. (Eds.). (2003). *Key Issues in Secondary Education*. London: Continuum.
Bernstein, B. (1971). On the Classification and Framing of Educational Knowledge. In M. Young (Ed.), *Knowledge and Control: New Directions for the Sociology of Education*. London: Macmillan.
Bernstein, B. (2000). *Pedagogy, Symbolic Control and Identity: Theory, Research and Critique* (Rev. ed.). London: Taylor and Francis.
Biesta, G. J. J. (2010). *Good Education in an Age of Measurement: Ethics—Politics—Democracy*. Boulder, CO: Paradigm.
Biesta, G. J. J. (2012). Giving Teaching Back to Education: Responding to the Disappearance of the Teacher. *Phenomenology and Practice, 6*(2), 35–49.
Bobbit, J. (1918). *The Curriculum*. Boston: Houghton Miffin.
Boyes, K., & Watts, G. (2009). *Developing Habits of Mind in Secondary Schools*. Heatherton, VIC: Hawker Brownlow.
British Government. (2018). Guidance: Protecting Children from Radicalisation: The Prevent Duty. Retrieved September 2018, from https://www.gov.uk/government/publications/protecting-children-from-radicalisation-the-prevent-duty
Bustin, R. (2018). What's Your View? Curriculum Ideologies and Their Impact in the Geography Classroom. *Teaching Geography, 43*(2), 61–63.
Carr, W. (2004). Philosophy and Education. *Journal of Philosophy of Education, 38*(1), 55–73.
Dawson, G. (1981). Objectivism and the Social Construction of Knowledge. *Philosophy, 56*(217), 414–423.
Donnelly, C. (2000). In Pursuit of School Ethos. *British Journal of Educational Studies, 48*(2), 134–154.
Dreeben, R. (1968). *On What is Learned in School*. Reading, MA: Addison-Wesley.
Firth, R. (2011). Making Geography Visible as an Object of Study in the Secondary School Curriculum. *Curriculum Journal, 22*(3), 289–316.
Fraziern, A., & Sadkerm, Y. (1975). *Sexism in School and Society*. New York: Harper & Row.
Furedi, F. (2007). Introduction: Politics, Politics, Politics. In R. Whelan (Ed.), *The Corruption of the Curriculum*. London: CIVITAS.
Gordon, D. (1982). The Concept of the Hidden Curriculum. *Journal of Philosophy of Education, 16*(2), 187–198.

Graves, N. (1975). *Curriculum Planning in Geography*. London: Heniemann.
Green, T. (1980). *Predicting the Behaviour of the Educational System*. Syracuse University Press.
Gutmann, A., & Thompson, D. (1996). *Democracy and Disagreement*. Cambridge University Press.
Harrison, J. L. (1977). Review Article: John Wilson as Moral Educator. *Journal of Moral Education*, 7(1), 50–63.
Haydon, G. (2000). John Wilson and the Place of Morality in Education. *Journal of Moral Education*, 29(3), 355–365.
Hirsch, E. D. (1988). *Cultural Literacy: What Every American Needs to Know*. New York: Random House.
Illichi, V. (1971). The Breakdown of Schools: A Problem or a Symptom? *Interchange*, 2, 1–10.
John, P., & Osborn, A. (1992). The Influence of School Ethos on Pupils' Citizenship Attitudes. *Journal of Educational Review*, 44(2), 153–165.
Lambert, D. (2016). Geography. In D. Wyse, L. Hayward, & J. Pandya (Eds.), *The Sage Handbook of Curriculum, Pedagogy and Assessment*. London: Sage Publications.
Lambert, D., Solem, M., & Tani, S. (2015). Achieving Human Potential through Geography Education: A Capabilities Approach to Curriculum Making in Schools. *Annals of the Association of American Geographers*, 105(4), 723–735.
Lovat, T. (1988). Curriculum Theory: The Oft-missing Link. *Journal of Education for Teaching*, 14(3), 205–213.
Marsden, W. (1989). All in a Good Cause: Geography, History, and the Politicization of the Curriculum in Nineteenth and Twentieth Century England. *Journal of Curriculum Studies*, 21, 509–526.
Maton, K. (2010). Canons and Progress in the Arts and Humanities: Knowers and Gaze. In K. Maton & R. Moore (Eds.), *Social Realism, Knowledge and the Sociology of Education: Coalitions of the Mind*. London: Continuum.
McLaughlin, T. (2003). Values in Education. In J. Beck & M. Earl (Eds.), *Key Issues in Secondary Education*. London: Continuum.
Moore, R., & Muller, J. (1999). The Discourse of 'Voice' and the Problem of Knowledge and Identity in the Sociology of Education. *British Journal of Sociology of Education*, 20, 189–206.
Morgan, J. (2015). Michael Young and the Politics of the School Curriculum. *British Journal of Educational Studies*, 63(1), 5–22.
Morgan, J., & Lambert, D. (2005). *Geography: Teaching School Subjects 11–19*. London: Routledge.

Osborn, M., Croll, P., Broadfoot, A., Pollard, E., McNess, E., & Triggs, P. (1997). Policy into Practice and Practice into Policy: Creative Mediation in the Primary Classroom. In G. Helsby & G. McCulloch (Eds.), *Teachers and the National Curriculum*. London: Cassell.
Priestley, M. (2011). Whatever Happened to Curriculum Theory? Critical Realism and Curriculum Change. *Pedagogy, Culture & Society, 19*(2), 221–237.
Pring, R. (2005). The Strengths and Limitations of 'Subjects'. *Nuffield Review of 14–19 Education and Training: Aims, Learning and Curriculum Series, Discussion Paper 1*.
Rawling, E. (2000). Ideology, Politics and Curriculum Change: Reflections on School Geography 2000. *Geography, 85*(3), 209–220.
Reiss, M., & White, J. (2013). *An Aims Based Curriculum*. London: IOE Press.
Roberts, M. (2003). *Learning through Enquiry: Making Sense of Geography in the Key Stage 3 Classroom*. Sheffield: Geographical Association.
Roberts, M. (2013a). *Geography through Enquiry: Approaches to Teaching and Learning in the Secondary School*. Sheffield: Geographical Association.
Roberts, R. (2013b). *Powerful Knowledge: A Critique*. Debate Given at the UCL Institute of Education, 13 May 2013. Retrieved April 2019, from https://www.youtube.com/watch?v=DyGwbPmim7o
Roberts, R. (2014). Powerful Knowledge and Geographical Education. *The Curriculum Journal, 25*(2), 187–209.
Schiro, M. (2007). *Curriculum Theory: Conflicting Visions and Enduring Concerns*. Thousand Oaks, CA: Sage Publications.
Scott, D. (2008). *Critical Essays on Major Curriculum Theorists* (2nd ed.). Abingdon: Routledge.
Silverman, A. (1970). *Crisis in the Classroom*. New York: Random House.
Standish, A. (2007). Geography Used to be About Maps. In R. Whelan (Ed.), *The Corruption of the Curriculum*. London: CIVITAS.
Standish, A. (2009). *Global Perspectives in the Geography Curriculum: Reviewing the Moral Case for Geography*. London: Routledge.
Stenhouse, L. (1975). *An Introduction to Curriculum Research and Development*. London: Heinemann.
Trigg, R. (1973). *Reason and Commitment*. Cambridge: University Press.
Trowler, P. R. (1998). *Academics Responding to Change: New Higher Education Frameworks and Academic Cultures*. Buckingham: Society for Research in Higher Education and Open University Press.
Tubb, C. (2003). Moral Education. In J. Beck & M. Earl (Eds.), *Key Issues in Secondary Education*. London: Continuum.

Usher, R., & Edwards, R. (1994). *Post Modernism and Education*. London: Routledge.
Wheelahan, L. (2010). Competency-based Training, Powerful Knowledge and the Working Class. In K. Maton & R. Moore (Eds.), *Social Realism: Knowledge and the Sociology of Education*. London: Continuum.
Whelan, R. (Ed.). (2007). *The Corruption of the Curriculum*. London: CIVITAS.
William, D. (2009). Assessment for Learning: Why, What and How? Inaugural Professorial Lecture, UCL Institute of Education.
Wilson, J. (1990). *A New Introduction to Moral Education*. London: Cassell.
Wilson, J., Williams, N., & Sugarman, B. (1967). *Introduction to Moral Education*. Harmondsworth: Penguin.
Young, M. (1971). *Knowledge and Control: New Directions for the Sociology of Education*. London: Collier-Macmillan.
Young, M. (2008). *Bringing Knowledge Back In: From Social Constructivism to Social Realism in the Sociology of Education*. Abingdon: Routledge.
Young, M., & Lambert, D. (2014). *Knowledge and the Future School: Curriculum and Social Justice*. London: Bloomsbury.
Young, M., & Muller, J. (2010). Three Educational Scenarios for the Future: Lessons from the Sociology of Knowledge. *European Journal of Education*, 45(1), 11–27.

3

Bringing the 'Geography' Back in

Introduction

The debates of the previous chapter outline why some writers have argued the curriculum has been in 'crisis' over the past few decades, especially relating to the confused role of knowledge in the curriculum. This chapter is specifically about geography as a school subject, but many of the ideas could relate to any school subject. The title of the chapter is based on Michael Young's (2008) work, entitled 'bringing knowledge back in'. It was this publication in which he argued for the inclusion of powerful knowledge, from subject disciplines, to be central to a curriculum. He is a former school Chemistry teacher so was able to relate many of his ideas back to his classroom days. This chapter is therefore 'bringing the geography' back into the debates around specialised knowledge, curriculum and Future 3 (F3) curriculum thinking.

The Specialised Knowledge of Geography

Geography is a form of 'specialised' knowledge, a term from Durkheim (e.g. Durkheim 1956). Specialised knowledge is that which is created and maintained by a subject discipline. This specialisation leads to a "focus on the shared values on which the objectivity of knowledge depends" (Young 2008, p. 208). Those shared values are held by academic researchers, working in universities around the world, who create new geographical knowledge within the 'rules' of the academic discipline of geography within which they work. It is the specialisation which enables 'better' knowledge to be created and as such creates 'socially realist' geographical knowledge (as discussed in Chap. 2).

'Geography' as a specialised knowledge has a long tradition. Translated literally from its Greek origins, the word 'geography' means 'earth description'; *geo* meaning 'earth', and *graphia* 'description'. Yet, as Unwin (1992) describes,

> geography is one of the oldest forms of intellectual enquiry, and yet there is little agreement among professional geographers as to what the discipline actually is, or even what it should be. (p. 1)

Geography has been described as the 'world discipline' (Bonnett 2008); a 'field of knowledge' (Walford 2000), a 'realm of meaning' and even a 'dimension of experience' (Livingstone 1992). For The Royal Geographical Society,

> geography is the study of the earth's landscapes, peoples, places and environments. It is, quite simply, about the world in which we live. (RGS IBG 2015)

Yet, Small and Witherick (1995) are quick to point out:

> [I]t is highly unlikely that any one definition of the subject would satisfy everyone…The fact that geography is located at the interface between the natural and social sciences adds to the difficulty in arriving at a definitive definition. (p. 100)

This gives geography a broad and ambitious definition. Yet this is problematic. The notion of a 'specialised' discipline suggests a coherence and all these articulations of geography suggest a discipline which is completely 'unspecialised'. Johnston (1997) expresses this clearly when he denies the existence of a separate discipline of geography at all. As he argues,

> to most of us, there is no such thing as geography, other than as a vaguely defined discipline to which we are attached as much for political and economic (that is, job security) reasons as for intellectual ones… And does it matter? I believe not. There is no such thing as geography, only a lot of separate geographies all of which share characteristics with the others, but are quite considerably self sufficient. (p. 35)

One characteristic of a coherent discipline is a similarity in how new knowledge is made. Yet this is problematic for geography. For Johnston (1997), far from there being one coherent set of rules underpinning knowledge creation in the discipline of geography, a multitude of methodologies exist depending on what is being studied.

The discipline is frequently classified in terms of both 'human' geography, the part of the subject that deals with human interaction, and 'physical' geography, the side that is concerned with the characteristics of the earth. Physical geography follows much of the specialist constructs of the physical sciences. Understanding coastal geomorphology, for example, requires a knowledge of determinable processes, such as wave erosion and how they impact on the landscape of the coastline. Yet human geography is much more akin to the humanities and sociological subjects with a multitude of methodologies and practices. Much of the knowledge in human geography is generated through opinions and understandings of phenomena. Understanding cultural geography of lived spaces, for example, requires the testimony of people living in places which geographers then make sense of. This means the traditional discipline has a schism right at the heart, which affects the means by which specialised geographical knowledge is created.

The split of geographical methodologies is evident to such an extent that Eden (2005) identifies, "few academics can now individually be both physical scientists and social scientists" (p. 285) and Furlong and Lawn (2011) identify "increased fragmentation, certainly between human and physical geography, but also within human geography" (p. 125). In academic geography the nature of geographical knowledge has been continually changing, and one of the more significant changes since the 1990s has been geographers working at the 'edges' of the disciplinary boundaries. As Massey (1999) explains,

> some of the most stimulating intellectual developments of recent years have come either from new, hybrid places (cultural studies might be an example) or from places where boundaries between disciplines have been constructively breached and new conversations have taken place. (p. 421)

The 'discipline' of geography is thus able to encompass a variety of methodologies, borrowing from a range of other discipline areas. Therefore the 'social realist' nature of geography and the rules that determine how new knowledge is created cannot be used to define what makes something geographical. Whether working with the sciences of physical geography or sociological approaches of human geography, *what* is being researched defines geography more than *how* it has been researched.

Yet defining what geographical knowledge is has proved equally as problematic as finding a unifying geographical methodology. One of the recurring main ideas in various descriptions of geography is the notion that it is the study of the earth as the home of the human race (e.g. Small and Witherick 1995). Its knowledge scope therefore is vast, as the Editorial from *The Times* newspaper on 7 June 1990 (the day the final report of the National Curriculum geography working group was published) describes:

> [G]eography embraces every fact on earth: every aspect of the composition, occupation and history of the planet… As such, geography holds no intellectual boundaries. (Reprinted in Boardman and McPartland 1993d, p. 146)

This broad scale definition proves equally unhelpful in terms of finding a unifying content for a social realist discipline. The knowledge content of geography has been expressed by a number of writers in terms of a series of 'key concepts' which underlie the nature of knowledge in the

Table 3.1 The key concepts of geography (Taylor 2009)

Leat (1998)	Geography Advisors' and Inspectors' Network (2002)		Rowley and Lewis (2003)
Cause and effect	Bias	Inequality	Describing and classifying
Classification	Causation	Interdependence	
Decision making	Change	Landscape	Diversity and wilderness
Development	Conflict	Location	Patterns and boundaries
Inequality	Development	Perception	Places
Location	Distribution	Region	Maps and communication
Planning	Environment	Scale	
Systems	Futures	Uncertainty	Sacredness and beauty
Holloway et al. (2003)	**UK 2008 Key Stage 3 curriculum (QCA 2007)**		**Jackson (2006)**
Landscape and environment	Cultural understanding and diversity		Proximity and distance
			Relational thinking
Physical systems	Environmental interaction and sustainable development		Scale and connection
Place			Space and place
Scale	Interdependence		
Social formations	Physical and human processes		
Space	Place		
Time	Space		
	Scale		

discipline. Taylor (2009), working in geography education, has collated these in Table 3.1.

Taylor's (2009) lists of concepts do reveal some common themes that get closer to the heart of the knowledge of the discipline; 'place', 'space' and 'scale' appear on multiple lists. 'Place' could be studied from a variety of perspectives; it has a physical and a human geography rooted in a location. Geographers are able to create knowledge of these concepts using a variety of methodologies.

One of the best articulations of the ways geographical knowledge can be specialised is from the field of geography education. Lambert (2004) suggests geography is a specialised 'language', and he differentiates between the 'vocabulary' and 'grammar' of geography. As Jackson (2006) continues,

(the)... *vocabulary*, (is) an apparently endless list of place names, and its *grammar*, (is) the concepts and theories that help us make sense of those places. (p. 199, original emphasis)

This 'grammar' is the various concepts that hold the pieces of knowledge (vocabulary) together, and this has been called 'thinking geographically' by Jackson (2006). It is this way of conceptualising geographical knowledge that resists the temptation of listing content, but by offering a set of principles about how geographical knowledge can be conceptualised, it gives the knowledge content its specialist status. Thinking geographically is being able to articulate the work of those geographers for whom the concepts and theories are related to physical processes such as erosion and deposition as well as those whose processes involve human migration and global flows of ideas.

Another means by which geography could be seen as a distinctive, social realist discipline is through the skills embedded within the subject discourse. Yet again, there is no distinctly 'geographical' skill; geographers use a variety of skills to enable knowledge development. One such skill is the use of spatial data and statistical information, and whilst the mathematical aspects of this would not be distinctly geographical, the accurate representation and interpretation of spatial information through cartography could be seen to be a geographical skill. Referring to the work of geography teachers, Lim (2005) argues:

> [A] challenge for all geography teachers, regardless of experience, has been to help students form in their minds a three-dimensional understanding of a given place, using only the information from a two-dimensional topographic map. (p. 187)

A map on its own, however, is not distinctly geographical as other subjects can use them in their discourse. It is how maps are used to enhance geographical knowledge and understanding that makes them geographical. Balchin and Colman (1971) described this skill as 'graphicacy', and the ability to represent, interpret and analyse information cartographically can be seen to be a geographical skill, and this includes both traditional and digital mapping such as Geographical Information Systems (GIS).

The significance of geographical knowledge, skills and thinking has been illustrated by Hulme (2008) in relation to the academic discourse on climate change. He was concerned by the lack of geographers contributing to debates at policy level. As he argued,

> the construction of narratives around global warming remain strongly tied to roots within the natural sciences... I am increasingly convinced that making human sense of climate change needs the distinctive intuition and skills of the geographer. These intuitions include long familiarity with working at the boundaries between nature and culture ... a tradition of understanding the subtleties of how knowledge, power and scale are inseparable ... We need new ways of thinking about and understanding the hybrid phenomenon of climate change. Geographers have a unique role to play in this task. (pp. 5–6)

For Hulme (2008), although climate change is a concept that crosses disciplines, geography has a unique role to play in developing understanding of the concept. He is convinced that there is a unique geographical knowledge content of climate change and his expressions of how geographers can contribute to the debate have close links to many of the concepts identified by Taylor (2009) and discussions of the interdisciplinary thinking of Massey (1999) and geographical thinking of Jackson (2006); scale, place and boundary thinking. It is through these concepts that geography can claim to be a specialised knowledge. The discipline of geography is itself 'ill- disciplined', but with a focus on key concepts such as place and thinking geographically, some unity in the discipline can be found.

Alastair Bonnett (2008) describes geography as "one of humanity's big ideas" (p. 2). He created a nine-stage definition of the discipline, which covers a wide range of ideas and concepts, arguing geography

> is rooted in the human need for survival, in the necessity of knowing and making sense of the resources and dangers of our human and physical environment... Geography is an attempt to both understand and meet the world. (p. 121)

Bonnet's aspirations for the importance of geographical knowledge are similar to the aspirations of powerful geographical knowledge, and the idea of 'capabilities' in this book. Geographical knowledge has an importance beyond the set of facts and ideas it describes and the subject in schools should be able to articulate this clearly.

Geography in Schools

If the National Curriculum itself was 'aimless' in its first incarnation as White (2006) asserts so too was the justification for the inclusion of geography as a subject in schools. As such over time, the purpose of the subject seems to have changed from a subject teaching young people valuable knowledge to one where the subject is simply a vehicle for other skills and competencies. The view that subject knowledge is unimportant is echoed by some teachers; in research into 'place' education, Fanghanel (2009) researched a geography teacher in higher education who,

> strongly de-emphasised the disciplinary input in her approach, stating that she didn't feel that her 'duty was to turn out geographers' and underplaying her own sense of belonging: 'I have no big disciplinary allegiance. I like geography because it allows me to do the things that I like doing'. For her, the link to the discipline was less important than a sense that her students should access a broader understanding of the social world through her input. (p. 112)

This teacher, despite being a geographer, does not see the value of geographical knowledge in the curriculum. Her child-centred ideological approach reduces the significance of knowledge, and her views see skills replace knowledge as the central concern of a curriculum. This ill-disciplined thinking cannot provide a coherent structure for students to make sense of the world; it simply enables them to access small pieces of information that have no coherence or underlying ideology. Her views may not be alone. It is this sort of teaching, often by non-specialist geography teachers, that may be responsible for the 'boring and irrelevant' geography teaching in the 2000s (e.g. Ofsted 2008, 2011). As a school subject, geography was enshrined in the National Curriculum of 1988 and in subsequent rewrites has retained its place as a subject, but

3 Bringing the 'Geography' Back in

continued curriculum pressures have seen its content and purpose change, as it adapted to various curriculum demands.

Figure 3.1 presents this as a crude timeline, showing the most significant changes to the discipline of geography in both schools and universities, based on the work of Walford (2000) and Boardman and McPartland (1993a, b, c, d).

Figure 3.1 shows a timeline of changes from pre-1960 through to the exam reforms of 2017. The top set of boxes shows the development of geography as an academic discipline in universities. It shows how the subject developed through a series of 'paradigms'; the relationship shown here suggests a linear progression but in fact these ideas are often developed concurrently; shown here is the main time they have become a dominant paradigm in the discourse. The 'regional approach', often dubbed the 'capes and bays approach' due to its descriptive nature of naming landforms and places, gave way to quantitative methodologies in the 1960s when the subject was using statistical methods to attempt to add rigour. This gave way to a swathe of more humanistic and behavioural approaches throughout the 1970s and 1980s which in turn influenced modernism and postmodernism in the 1990s and beyond (e.g. Walford 2000). This suggests geography is a vibrant and dynamic academic discipline, constantly evolving to help explain a changing world.

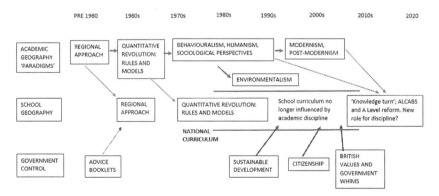

Fig. 3.1 A timeline of the changing nature of geography as an academic discipline and school subject (based on Walford 2000; Boardman and McPartland 1993a, b, c, d)

The second row of boxes maps out what was going on in schools at the same time. For much of the time shown, school geography seems to mirror academic geography, but with a 20-year time lag. Ideas from academic geography informed a generation of graduates but these were not incorporated into school curricula until those graduates were in influential positions in schools and as textbook writers and examiners. This relationship between academic and school geography did not last long as the National Curriculum of 1988 'fixed' the content of geography for the next 20 years. Thus school geography of the 2000s was still influenced by many of the advances made in the quantitative revolution of the 1960s and seemed untouched by many of the later humanist and modernist paradigms (e.g. Boardman and McPartland 1993a, b, c, d). This has meant the geographical knowledge component of school and university geography has been drifting apart in what Goudie (1993) described as the "great divide".

I was acutely aware of this myself as a young geography teacher. At university, I had been fascinated by postmodern urban geographies but in schools I was now teaching, as fact, urban land use models, simplified maps of a cityscape that were developed in the 1930s through to the 1960s to help describe cities. I found this a frustration; cities are fantastic places and geographers have unique ways to help understand the landscape but somehow teaching land use models as facts to be learnt was too simplistic. The lack of complexity was part of the reason geographers moved on from models to find new ways of understanding. Smith and Ogden's (1977) observation that "students entering university are often unprepared for the kind of geography that awaits them" (p. 47) was even more acute in the 2000s. It was this frustration that inspired me to research the 'great divide' as part of my master's level research. In this research (Bustin 2011a, b), I taught postmodern urban geographies, based on the work of Soja (1996), and 'Thirdspace' to secondary school children as part of their urban studies geography course to attempt to bridge this divide. Other teachers and writers were also drawing on contemporary geographical ideas to create inspiring classroom activities (e.g. Oakes 2004) but these were often small scale and piecemeal.

The final part of Fig. 3.1 shows the influence of the national government. In the 1960s and 1970s, advice to teachers came in the form of advice booklets (e.g. HMI 1978), but with the introduction of the National Curriculum the control becomes even tighter. In recent decades, the geography taught in schools has become increasingly influenced by government curricular changes. Geography has increasingly become a vehicle for education for sustainable development, citizenship and other political 'projects'. Part of the reason the subject seems to be able to embrace curricular changes is due to its vast potential knowledge base. This has always meant the subject knowledge base has been 'malleable' and has allowed various 'fashionable causes' (as described by Furedi 2007) to infiltrate and dominate the geography curriculum. It is this increasing dominance of the 'social education' component of the geography curriculum that unbalances what Marsden (1997) described as his idealised geography curriculum. For him, there are three parts of a geography curriculum which should be kept in balance. These are the *subject* component, *educational* component and *social education* component. As he warns,

> unhealthy stresses arise when the three basic components of curriculum planning are not kept in reasonable balance. One imbalance occurs if the *subject component* is given too high a priority, resulting in a domination of content…the second problem emerges when the *educational component* is over- stressed… the third tension arises when the *social education component*, often associated with a contemporary good cause or issue, holds sway. (Original emphases p. 242)

In particular, during the 2000s, the importance of responding to climate change and other environmental causes, pro-European sentiment and buying Fairtrade products all seemed to be promoted through the geography curriculum. Thus by the 2010s, much had been written about the geography curriculum's political causes, with an article in *The Times* newspaper entitled "School children are victims of a green conspiracy" (6 October 1997) followed on by an article in *The Independent* "Is geography brainwashing?" (6 February 2003). The ideas were explored more fully in two important books at the time. The first, *The Corruption of the Curriculum* (Whelan 2007), took a holistic view on the whole curriculum, and Standish's (2007) chapter 'Geography used to be about maps' set out the case for the overt politicisation of school geography. The second book, *Global Perspectives in the Geography Curriculum* (Standish 2009), is a continuation of these ideas. Standish (2009) illustrates the issue with regards to the teaching of 'Fairtrade' in geography lessons of the 2000s:

> [P]upils are not necessarily told what to think but the information presented is unlikely to lead one to question… the issue is presented to the students in simplistic, narrow and personal terms. There is no evaluation…. the issue has been removed from its wider social and political context making it solely a matter of individual consciousness. (p. 45)

Knowledge of Fairtrade and global consumption patterns has been removed. What Standish (2009) calls for is the teaching of Fairtrade to occur within a broader framework of geographical knowledge. He argues that only by understanding what fair trade, and therefore presumably unfair trade, actually is, and how it develops, can children engage with the concept. By increasing the knowledge basis and introducing children to ideas about globalisation, trade patterns and global economics, children can gain an understanding of these issues; they are then in a much stronger position to form their own values and opinions about their consumption patterns, rather than being told what to do and how to behave by their geography teachers, which was the accusation made by the *Times* and *The Independent* through those articles.

Whilst Standish (2007, 2009) was an advocate of an increased knowledge base for school geography, the type of 'core knowledge' Standish (2007, 2009) was promoting appeared to me to be similar to the ideas of Hirsch (1988)—that there was a set of core geographical knowledge that can define the content of school geography. It was this belief that set Standish apart from his peers. As Lambert and Morgan (2009) explain,

> Standish writes almost with a kind of wistful sense of loss, using the title 'geography used to be about maps' with only the slightest irony. His response to the particular concern of political interference is to laud the subject itself, as if subject knowledge *itself* were natural, stable and legitimate, and not a human creation subject to change… he argues for the importance of the body of knowledge in its own right and the need to pass this tradition on to young people. (p. 154)

Trying to define what knowledge might make up the 'core knowledge' of geography became the topic of debate amongst teachers, academics and subject communities at the end of the 2000s around the time of National Curricular reform. Standish put forward his version of a core knowledge-based curriculum, and this was considered alongside proposals from the Geographical Association (GA 2011)—the subject association for geography teachers—and the Royal Geographical Society (with the Institute of British Geographers). The knowledge content of the secondary geography curriculum was an important topical debate in geography education at this time, but was dominated by definitions of the subject and expressions of what core geographical knowledge actually was rather than trying to look more widely at why we want children to learn geography and some of the broader questions about the aims of education. Discussions ended up in further disagreement and confusion.

The final section of the diagram in Fig. 3.1 illustrates the current situation; the tight grip on defining content was reduced to a series of 'topics' in the 2010 reforms and teachers were able to interpret these for themselves. Yet without engaging critically with curriculum content for over 20 years, teachers would have found this a challenge. Whole scale rewrites of the GCSE (General Certificate of Secondary Education) and A Level courses occurred in 2015 and 2016. The content of these new A Level courses was influenced by the newly created ALCABS, 'A Level content advisory boards'. These new courses were very detailed on content to be learnt and re-established the close link between school and university geography with an introduction of topics such as 'place' and ideas from cultural geography which had been developed in the academic discipline from the 1990s. Yet this 'knowledge turn' seemed to yet again promote an approach to knowledge in which children learn page after page of facts rather than developing a critical understanding of what was being learnt.

Relating the changes to the geography curriculum back to the work of Young and Muller (2010), school geography can be expressed through both Future 1 (F1) and Future 2 (F2) visions. An F1 geography curriculum would be concerned with lists of facts and content to be learnt uncritically. Early versions of the National Curriculum listed content for teachers to teach in schools and if done uncritically this creates an F1 curriculum. A professional discussion amongst geography educators in the

2010s sought to define geography's 'core knowledge' and most attempts listed topics and facts to be learnt. This was Lambert and Morgan's (2009) criticism of Standish's proposal for curricular reform. The 'capes and bays' approach to school geography simply saw learning of facts and this image of school geography still persists in the popular imagination of the subject amongst many non-geographers. Conversely, an F2 curriculum sees geographical knowledge reduced and an increased emphasis on generic educational, but not 'geographical', skills, derived from the needs of governments such as healthy eating and developing empathy. When cut off from input from the academic discipline school geography became much more F2 orientated, with increased 'politicalisation' (e.g. Standish 2007).

The reforms of the 2010s and the influence of the ALCABS seem to have swung the curriculum pendulum back to F1 curriculum thinking, which could result in a dominance of content to be learnt.

The Future 3 Geography Curriculum and Powerful Geographical Knowledge

In a similar vein to discussions around the whole school curriculum, school geography seems to be presented through either an F1 or F2 curricular vision. Young and Muller's Future 3 curriculum could be an important idea to express school geography, drawing on ideas of powerful 'geographical' knowledge. Yet trying to explore what an F3 geography curriculum might be like presents challenges.

The first major challenge comes through the nature of geographical knowledge, and the relationship between the sort of knowledge created in universities and that experienced in school classrooms. Firth (2011) explored this relationship between the university disciplines and the associated school subject, suggesting "social realist theory seems to imply that.... the discipline inevitably precedes and delimits the school subject" (p. 305). Yet despite its classical origins, geography as an organised subject in schools predates the formal university discipline, with the latter starting to ensure a supply of well-qualified geography teachers (Young and Lambert 2014). If specialised knowledge is created by subject communities of experts, there is an immediate problem in identifying who these experts are in relation to geography; academics in universities or teachers in schools. Figure 3.1 modelled the often disconnected relationship, the 'great divide' (Goudie 1993) between geography as a school subject and as an academic discipline.

A challenge in creating an F3 geography curriculum comes through the nature of the 'classification' and 'framing' of the specialised knowledge of geography, and the curriculum of which geography is part, based on ideas from Bernstein (1973) introduced in the last chapter. Knowledge 'framing' identifies the ways knowledge is selected and bounded to form a discrete set of curriculum 'content', and allied to this is the notion of knowledge 'classification', which refers to "the degree of boundary maintenance between (these) contents" (Bernstein 1973, p. 205). In a curriculum that is 'strongly' classified, knowledge is bounded in discrete subjects that have no crossover or integration with each other. Conversely, a 'weakly' classified curriculum is one in which there is a large amount of crossover between traditional subject areas, and where knowledge is integrated, perhaps within broader topics or themes.

Figure 3.2 illustrates the nature of framing of geographical knowledge within a whole school curriculum diagrammatically.

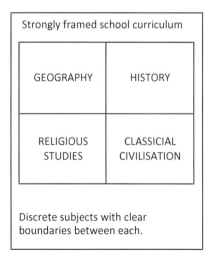
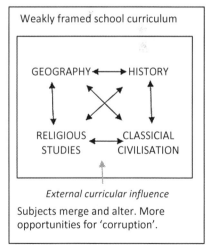

Fig. 3.2 The nature of curriculum framing. The weakly framed curriculum illustrated here sees geographical knowledge as part of a 'Humanities' subject, in which it loses its geographical identity

Classification and framing of knowledge can be combined to outline not only the challenges facing geography in schools, but also the changing importance of a subject-based school curriculum. Table 3.2 attempts to explain this relationship with the addition of plus and minus signs after the letters C for classification and F for framing, which were devised by Daniels (1987) to indicate the relative strength of the classification and framing (+ for strong and − for weak).

Table 3.2 The relationship between classification and framing of the geography school curriculum based on Bernstein (1973) and Daniels (1987)

	F++	F−−
C++	**Strong classification and strong framing**: Indicative of the National Curriculum for secondary schools of 1991 in which discrete subject disciplines existed to deliver a pre-determined set of content with national examinations. For geography, the subject would contain lists of highly prescriptive 'core' knowledge likely to incite an F1 curriculum.	**Strong classification and weak framing**: Pre-National Curriculum school curricular and post-2008 reforms of the National Curriculum provide a weak frame, by reducing specified content and returning more control to teachers, though still within subject disciplines. Geography teachers are able to choose their own content in geography lessons.
C−−	**Weak classification but strong framing**. This could be indicative of the experience of primary schools. Teachers are not bound by distinct subject disciplines, and those that do exist such as maths and English have close links. Teachers have more autonomy over the topics they teach but do have predetermined competencies and skills they need to deliver such as literacy, numeracy and IT skills. Geography is unlikely to be a discreet subject, but 'world knowledge' and some generic skills such as map interpretation would be integrated within a topic-based curriculum.	**Weak classification and weak framing**. This would be indicative of a school not bound by traditional subjects or any pedagogic expectation. Teachers would have complete autonomy in the classroom. Harley (2010) suggests in the US school curricula "both classification and framing have been historically weak" (p. 8),and in the English system some new schools now seem to be developing this form of curriculum, where geography is either non-existent or combined as part of 'humanities'.

The notions of classification and framing help to position many of the observations of the geography curriculum. For some schools free from government curricular control, the classification of subject boundaries is weak where some subjects like geography are completely lost from the curriculum or combined to form 'humanities'. The resultant curriculum, free from the 'constraints' of subjects tends towards an F2, skills- and competency-based curriculum. On the contrary, a strongly classified and highly framed curriculum has a rigid prescription of content, which tends towards an F1 curriculum interpretation in which 'core' knowledge has to be 'delivered'. A strongly classified, but relatively weakly framed, curriculum opens up possibilities for an F3 curriculum; classified to enable subjects to exist within a curriculum but weakly framed enough to encourage geography teachers to select knowledge for themselves and to not see content as something to 'deliver'. This type of curriculum structure exists in schools today and as such offers opportunities to develop an F3 curriculum.

A key component of an F3 curriculum is its emphasis on powerful knowledge. For teachers of geography in schools, presenting a coherent discipline for pupils becomes a challenge given its vast potential knowledge base. Defining the powerful knowledge of geography that needs to be taught in schools becomes as much of a challenge as delimiting the knowledge in the academic discipline. Any attempt to define powerful geographical knowledge creates a tick list of content (e.g. Hirsch 1988) and immediately suggests an F1 curriculum. Despite this, attempts have been made to identify powerful geographical knowledge; that is knowledge that has been created within the subject of geography by subject experts; is open to debate and discussion; represents the best that has been thought of in the subject and that moves beyond the 'everyday' into complex thoughts and ideas. Taylor's (2009) list of core concepts, Table 3.1, simply groups and classifies what has traditionally been part of prescriptive geographical content of the past.

Identifying the powerful knowledge of geography is problematic in part due to the 'horizontal structure' of geographical knowledge, which Bernstein (1996) identifies, and which he differentiates from 'hierarchical' knowledge structures. Horizontal knowledge structures are "a series of specialised languages, each with its own specialised modes of interrogation and specialised criteria" (Bernstein 1996, pp. 172–173).

Subject knowledge here is described as a series of distinct ideas, a way of understanding and communicating a phenomenon. It is characteristic of a weakly framed curriculum and would characterise subjects such as art and history. These subjects can be accessed at a higher level without a gaining of the lower level knowledge. A student could study a series of geographical topics in any order; there is no reason why a topic on 'coasts' should follow on from a topic on 'rivers'; they are distinct horizontally structured areas of geographical knowledge often with their own specialised 'language' of knowledge creation. In the same vein, a pupil could start a geography A Level without necessarily studying geography at GCSE.

By contrast, hierarchical (or vertical) knowledge structures are "an explicit, coherent, systematically principled and hierarchical organisation of knowledge, which develops through the integration of knowledge at lower levels" (Ibid., pp. 172–173). This type of organisation would see a progression in complexity of knowledge towards more abstract concepts, and so access to later knowledge relies on the gaining of earlier knowledge. In schools, this could be characterised by subjects such as maths and physics. A student needs to learn to multiply and divide before they can access algebra. A student could not start an A Level in maths without a GCSE in the subject. Figure 3.3 models these two types of knowledge structure. Vernon (2016) models vertical structures as a triangle, showing increased levels of complexity, and horizontal structures as interconnected circles. Here they are illustrated in rectangular blocks with illustrative examples drawn from maths and geography.

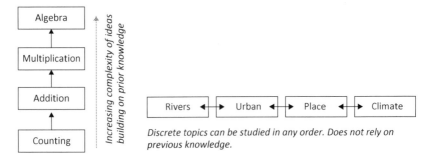

Fig. 3.3 Diagrammatic representation of vertical (left) and horizontal (right) knowledge structures (after Bernstein 1996)

The simple arranging of subject knowledge into either vertical or horizontal knowledge structures is over simplified. Within subjects with more horizontal structures, there will still be a degree of verticality. Some theories and concepts are common across various topics and develop the more that are studied; in geography, this could be ideas such as cycles of erosion and deposition being common amongst many otherwise separate topics, or the ability to interpret a map. Similarly, in otherwise highly structured subjects with more verticality, some topics and ideas can be explored in any order; learning about planets in physics, for example, does not need to follow on from studying light.

Identifying what might be considered powerful knowledge in subject disciplines with a more hierarchical knowledge structure might be an easier challenge due to the progressive nature of the subject. Yet for geography, this is not such a simple task.

The initial work on defining powerful geographical knowledge inevitably had to take a different approach than simply trying to 'define' lists of content. Instead, geography teachers were asked to think about something they already taught in the classroom to their pupils and to consider why this could be considered powerful knowledge. This work has been part of the GeoCapabilities 2 project and the topic of associated project workshops. In an attempt to provide a real-life example of what powerful knowledge in geography might look like, and how it differs from simply thinking in terms of a series of facts, Table 3.3 is a short 'reflective piece' developed at the start of the project (subsequently published in Bustin 2017). This piece explores the notion of powerful geographical knowledge in relation to the teaching of coastal geomorphology with a class.

Table 3.3, suggests that powerful knowledge positions information within its subject context, using the notions of 'thinking geographically' to express geographical knowledge. What the vignette is able to do is to enable a geography teacher to really consider the nature of what they are teaching and what makes it 'geographical', as opposed to history, physics or some 'corrupting' influence on the curriculum. Part of the GeoCapabilities 2 project encouraged teachers to create their own vignettes of powerful knowledge, versions of Table 3.3, to explore this notion further. There are many examples of these for many areas of school geographical knowledge and these can now be found on the project website (geocapabilities.org 2018).

Table 3.3 A 'vignette' of powerful geographical knowledge based on the teaching of coastal geomorphology in the geography classroom

DESCRIPTION
My Year 10 class (15 year olds) knows that 'the Holderness coastline (on the east coast of England) is made from boulder clay'.
This is not everyday knowledge. But is it 'powerful knowledge'? I would argue it is not, on its own, powerful knowledge. It is just a more or less correct 'fact'.
However, 'boulder clay' (or more precisely glacial till) could be conceptualised in a number of ways by different academic disciplines—chemists would be interested in its chemical composition, physicists might look at its tensile strength and the way it behaves under different stress and pressures. Geographers could look at it in a number of ways: for example, geomorphologists would develop their knowledge of boulder clay in terms of its physical properties of permeability, its tendency to slump and move under gravity and how it affects and is affected by its environmental context. To fully understand boulder clay geographically it needs to be placed within the context of its origins (from glacial deposition some 10,000–20,000 years ago) and its surroundings, which in the case of the clay on the Holderness coastline includes its location next to the sea. The actions of the sea (which can also be conceptualised in a number of ways) are of importance to understand the significance of the boulder clay as the wave action erodes the clay cliffs to cause rapid cliff retreat.
DISCUSSION
Our knowledge of 'boulder clay' (or glacial till) is shaped by the way it is conceptualised in the discipline of geography. For instance, we do not fully comprehend the significance of this phenomenon without knowledge of its origins, composition and location. It is this that makes it 'powerful'. It is almost the 'back story' of boulder clay—the way boulder clay is understood—that is indicative of the way geographers identify and describe it, and its significance.

The powerful knowledge vignettes created as part of the GeoCapabilities 2 project enabled geography teachers to think deeply about what they choose to teach their students in the geography classroom. It got them to engage with curriculum thinking. Focussing on powerful geographical knowledge for teachers ensures that there is a geographical content in lesson planning, answering the 'where's the geography?' question that Roberts (2010) asks of trainee teachers. Powerful knowledge is at the heart of an F3 curriculum, and so engaging with this moves closer towards an F3 geography curriculum. The discussions of powerful knowledge presented thus far rely on geography teachers, as specialists, being able to identify and justify the geographical content of their lessons. It is a discussion about 'what' to teach, and what should be an input to a school geography curriculum.

Yet there is a much larger debate about 'why' this particular knowledge has been selected to be taught, relating back to the ideas of ideologies and aims of education. If Bonnett (2008) is right that geography is "one of humanity's big ideas" then powerful knowledge should be more ambitious in its definition and scope, putting the emphasis back to the ways in which the knowledge itself can be empowering. As a response to this, later work on powerful knowledge has changed emphasis. Rather than trying to define or express powerful knowledge, more recent work has asked in what ways geographical knowledge can be considered 'powerful' for young people. The focus therefore becomes more about the young people being educated themselves and their curricular outcomes. This idea is explored in the work of Maude (2016) who has expressed five 'types' of geographical knowledge that give 'power' to school students. These are outlined in Table 3.4.

Maude's (2016) types of knowledge move beyond many of the descriptions of geographical knowledge that have gone before (such as Taylor's key concepts), and the various vignettes illustrated by Table 3.3, by highlighting the ways in which geographical knowledge can enable pupils to think in new ways, and participate in geographical debates and discussions. Maude's 'types' are purposefully broad and gives teachers freedom to choose geographical content for themselves for their pupils to engage with. There are links between these types of knowledge and some of the key concepts of Taylor (2009) with 'significant local, national and global

Table 3.4 Types of powerful knowledge in geography (Maude 2016)

Type 1	Knowledge that provides students with 'new ways of thinking about the world'.
Type 2	Knowledge that provides students with powerful ways to analyse, explain and understand the world.
Type 3	Knowledge that gives students some power over their own knowledge.
Type 4	Knowledge that enables young people to follow and participate in debates on significant local, national and global issues.
Type 5	Knowledge of the world.

issues' being a part of what Taylor might express as 'scale'. Reflecting on the vignette created in Table 3.3, the power of this knowledge is related to Maude's type 2 knowledge; by learning about the nature of an eroding cliff line, pupils gain knowledge of the sea, the land and the weather in new ways, by understanding how these concepts are interrelated.

Yet some of these types of knowledge are not solely related to geography. The word 'world' recognises the spatial element of this knowledge but type 3, which recognises the ability of pupils to question sources of knowledge, could be an element of the powerful knowledge of potentially all school subjects. Maude's (2016) list cannot solely be considered examples of powerful 'geographical' knowledge. Defining types of powerful knowledge in this way is vague on specific content, unlike the vignettes and teacher reflection which enabled more focus as they are grounded in classroom experience. On their own, neither Maude's typology, nor the vignettes of powerful knowledge, nor the other ways of defining geographical knowledge such as the key concepts enable a meaningful and engaging curriculum, and not yet an F3 curriculum. A more holistic framework of ideas is needed in order to take this thinking further and towards F3 curriculum thinking. This is where 'curriculum making' can be a useful concept.

The 'Making' of a Future 3 Geography Curriculum

Rethinking how the knowledge presented to students in a classroom might be considered powerful knowledge, rather than seeing it as facts to deliver, increases the importance of subject knowledge and the significance of what makes it distinctly 'geographical'. It is powerful knowledge, rather than facts, at the heart of a curriculum that is indicative of Future 3 curriculum thinking. Teachers can use their understanding of geography to help children to engage with the powerful knowledge of the subject, a process developed with the Geographical Association which Lambert and Morgan (2010) call 'curriculum making'. Curriculum making articulates the process of linking powerful geographical knowledge to pedagogy (what Roberts 2014 could consider 'powerful pedagogies'). It is allied to discussions of recontextualisation of knowledge (from Bernstein 2000) but takes this further by including considerations of the pupils and the school setting. It is this that has the possibility of creating F3 curriculum thinking.

Curriculum 'making' can trace its origins back to the work of Bobbitt (1918), who argues a school curriculum is something that needs to take children beyond their everyday knowledge and experience (as discussed in Catling 2013). Curriculum making is not lesson planning, but occurs over much longer timescales than one or two lessons by looking holistically at what, why and how pupils learn throughout a course of study in a subject. Curriculum making is not course planning, curriculum development or interpreting an exam specification as these can take place without a teacher or a pupil involved in the process.

Since its inception in the early twentieth century, the complex set ideas that helps shed light on the interaction of teachers, pupils and subject knowledge has developed across a variety of perspectives, with the German/Nordic 'Didaktik tradition' and the French 'subject didactics' or 'didactiques des disciplines' (e.g. Hudson 2016). Developed from this thinking is the 'didactic triangle', a model that helps to map out three considerations that exist when thinking about curriculum; the student; the teacher and the 'content', what could be called powerful knowledge. These three ideas make up the vertices of the triangle and the sides which connect them explore the relationship between the ideas. Linking the teacher to the student is the idea of pedagogical interaction, how teaching is being done. Linking the content to the student is experience and how the students respond to and make sense of the content. Linking the teacher to the content is the idea of representation of knowledge, what Bernstein would refer to as 'recontextualisation' (2000). The basic triangle sits within a broader set of ideas about school and curricular values; as Gericke et al. (2018) assert, "the didactic triangle is socially and culturally embedded" (p. 437). Many versions of the didactic triangle have been drawn and presented (e.g. Klafki 1997) showing the various elements interacting. 'Curriculum making' can be seen as an Anglo-Saxon interpretation of the same set of ideas, developed within geography education by the Geographical Association and Lambert and Morgan (2010).

To achieve successful curriculum making, for Lambert and Morgan (2010), three considerations should be kept in balance: the powerful knowledge of geography as a subject discipline; powerful pedagogies or teachers' choices about how to enable children to engage with geography; and the students themselves, their experiences of the world and the 'everyday knowledge' they bring to the classroom, their motivations and the ways they learn. These are similar ideals to those in the didactic triangle. This all takes place within the context of the discipline of geography, where new geographical knowledge is created, argued over and presented. Figure 3.4 models the process of 'curriculum making' in geography, amended to include ideas of powerful knowledge and powerful pedagogies.

The ideal curriculum for Lambert and Morgan (2010) is at the centre of the conceptual diagram, where all three circles combine. As they argue:

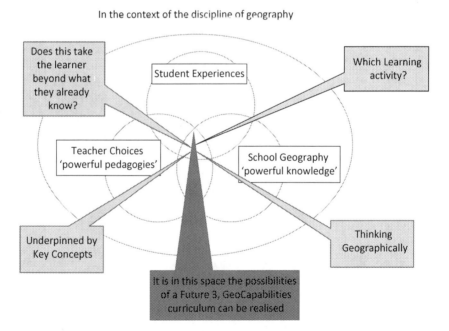

Fig. 3.4 A model of 'curriculum making in geography' (based on Lambert and Morgan 2010, p. 50)

If the subject was the only real source of creative energy (and students were thought to be passive recipients of what the teacher, literally, had to tell them) then we return to the often caricatured grammar school ways of the mid twentieth century. On the other hand, if the curriculum is made entirely to serve students interests and is entirely 'child centred' it risks taking them nowhere new in their learning and could be as deeply unproductive as the previous option. Finally, if the teachers were only really interested in their own performance and the 'pedagogical adventure' they can provide, the curriculum risks being emptied of meaning, overtly driven by skills at the expense of knowledge and understanding. (p. 51)

This discussion is a natural extension of Marsden's (1997) work on a 'balanced curriculum', with the subject, educational and social education aspects needing to be kept in balance. The discussion can also link to Young and Muller's (2010) Futures curriculum with F1 being an overemphasis on knowledge and F2 being an over focus on teachers' performance or children's needs. The ideal curriculum is one in the centre, where all parts are balanced, and teachers are able to mediate between the demands of knowledge, of pupil needs and of pedagogy. It is in this space that envisioning an F3 curriculum becomes possible.

Achieving successful curriculum making in practice requires teachers to consider their role in more ambitious terms than simply passing on a pre-determined body of knowledge or preparing students to take an exam. Brooks (2016) has explored this idea in terms of teacher professional 'identity' and the cultures within which they work. Her research studied a group of geography teachers over an extended period of time, and how their attitudes towards and relationship with their subject changed as their life circumstances and career progressed. Her work helps us to understand the changing attitude and motivations of teachers as professionals taking on the role of curriculum making.

Teachers need to take the responsibility for curriculum making. It is teachers who are the subject specialists working with a deep understanding of their subject, which enables powerful knowledge to be at the heart of their teaching. Teachers need to ensure they are at the forefront of developments in their discipline and stay updated throughout their career. They need to stay on top of pedagogical innovation to help their students engage in a meaningful way with powerful knowledge. They need to keep updated with the aspirations and needs of their students to ensure a dynamic and relevant curriculum. This is why the GeoCapabilities 2 project team used the subtitle 'teachers as curriculum leaders' to describe the attitude and approach needed from teachers to develop Future 3 thinking. Teachers are the leaders in the classroom, leading on curriculum and leading on pedagogy. Brooks (2016) identifies this attitude as a teacher's professional compass, navigating through the complexities of changing educational landscapes. It is through this raised professional responsibility that an F3 curriculum can be realised.

Conclusions

This chapter is solely about geography, 'bringing the geography back in' to the challenges and ideas set up in Chaps. 1 and 2. This is not straightforward given the challenge of defining what constitutes geographical knowledge both in the academic discipline and in schools. Effective curriculum making embeds powerful geographical knowledge within a broader set of considerations for teachers to 'make' the geography curriculum. Powerful knowledge helps to focus teachers' minds on the important curriculum question about 'what' is being taught, and from this the pedagogical questions about how to teach it can be considered. By reframing the debate around powerful knowledge, from attempting to define it as a list of content to considering the ways in which knowledge is 'powerful' for young people, we can start to link knowledge in schools to wider educational aspirations. Understanding how knowledge could be considered 'powerful' begins to answer the question about 'why' we choose to teach that knowledge in the first place. The idea that the true 'power' of geographical knowledge is articulated by the ways in which the knowledge is beneficial for young people is one of 'humanities' big ideas'. This is close to the aspirations of 'capabilities' and this is explored further in the next chapter.

Questions to Consider:
1. What makes geography a form of specialised knowledge?
2. What should the relationship be between geography as an academic discipline and as a school subject?
3. What sorts of classroom activities would be indicative of F1 and F2 curriculum thinking? How would F3 be different?
4. How can effective 'curriculum making' be of benefit to teachers?
5. Create a 'vignette' to illustrate the powerful knowledge of geography, similar to Table 3.2. How does this illustrate the powerful knowledge of geography? You can compare these with some examples on the geocapabilities.org website.
6. How do the arguments presented here relate to 'world knowledge' encountered by students in schools where geography is not part of the school curriculum? Do students still have access to powerful geographical knowledge?

References

Balchin, W., & Colman. (1971). Graphicacy; The Fourth Ace in the Pack. *Geography, 57*(3), 185–195.
Bernstein, B. (1973). *Class, Codes and Control* (Vol. 1). London: Routledge.
Bernstein, B. (1996). *Pedagogy, Identity and Control*. Lanham: Rowman & Littlefield.
Bernstein, B. (2000). *Pedagogy, Symbolic Control and Identity: Theory, Research and Critique* (Rev. ed.). London: Taylor and Francis.
Boardman, D., & McPartland, M. (1993a). Building on the Foundations: 1893–1943. *Teaching Geography, 18*(1), 3–6.
Boardman, D., & McPartland, M. (1993b). From Regions to Models: 1944 to 1969. *Teaching Geography, 18*(2), 65–68.
Boardman, D., & McPartland, M. (1993c). Innovation and Change: 1970–1982. *Teaching Geography, 18*(3), 117–120.
Boardman, D., & McPartland, M. (1993d). Towards Centralisation. *Teaching Geography, 18*(4), 159–163.
Bobbitt, J. (1918). *The Curriculum*. Boston: Houghton Mifflin.
Bonnett, A. (2008). *What is Geography?* London: Sage.
Brooks, C. (2016). *Teacher Subject Identity in Professional Practice: Teaching with a Professional Compass*. Abingdon: Routledge.

Bustin, R. (2011a). The Living City: Thirdspace and the Contemporary Geography Curriculum. *Geography, 96*(2), 61–62.
Bustin, R. (2011b). Thirdspace: Exploring the 'Lived Space' of Cultural 'Others'. *Teaching Geography, 36*(2), 55–57.
Bustin, R. (2017). *GeoCapabilities: Exploring the Powerful Disciplinary Knowledge of Geography*. Notes and Queries, The Newsletter of the Independent School Special Interest Group of the Geographical Association. Retrieved from December 2018, from https://www.geography.org.uk/write/MediaUploads/Get%20involved/GA_ISSIGNQ17.pdf
Catling, S. (2013). Teachers' Perspectives on Curriculum Making in Primary Geography in England. *The Curriculum Journal, 24*(3), 427–453.
Daniels, H. R. J. (1987). *An Enquiry into Different Forms of Special School Organization, Pedagogic Practice and Pupil Discrimination*. Unpublished doctoral dissertation, University of London, London.
Durkheim, E. (1956). *Education and Sociology*. London: Macmillan.
Eden, S. (2005). Commentary: Green, Gold and Grey Geography: Legitimating Academic and Policy Expertise. *Transactions of the Institute of British Geographers, 30*, 282–286.
Fanghanel, J. (2009). *Pedagogical Constructs: Socio-cultural Conceptions of Teaching and Learning in Higher Education*. VDM Verlag.
Firth, R. (2011). Making Geography Visible as an Object of Study in the Secondary School Curriculum. *Curriculum Journal, 22*(3), 289–316.
Furedi, F. (2007). Introduction: Politics, Politics, Politics. In R. Whelan (Ed.), *The Corruption of the Curriculum*. London: CIVITAS.
Furlong, J., & Lawn, L. (Eds.). (2011). *Disciplines of Education: Their Role in the Future of Educational Research*. Routledge.
Geographical Association. (2011). *The Geography National Curriculum: GA Curriculum Proposals and Rationale*. Retrieved July 2, 2013, from www.geography.org.uk/download/ga_gigcccurriculumproposals.pdf
Geography Advisors and Inspectors Network. (2002). *Thinking about the Future*. Unpublished Handout.
Gericke, N., Hudson, B., Olin-Scheller, C., & Stolare, M. (2018). Powerful Knowledge, Transformations and the Need for Empirical Studies across School Subjects. *London Review of Education, 16*(3), 428–444.
Goudie, A. (1993). Schools and Universities—The Great Divide. *Geography, 78*(4), 338–339.
Harley, K. (2010). *Draft Outline of Bernstein's Concepts*. South African Institute for Distance Education. Retrieved July 2019, from http://www.saide.org.za/resources/Library/Draft%20outline%20of%20Bernstein's%20concepts.doc

Hirsch, E. D. (1988). *Cultural Literacy: What Every American Needs to Know*. New York: Random House.
HMI Inspectorate. (1978). *Curriculum 11–16: Geography*. A working paper by the Geography Committee of HM Inspectorate, Department of Education and Science, London.
Holloway, S., Rice, S., & Valentine, G. (2003). *Key Concepts in Geography*. London: Sage.
Hudson, B. (2016). Didactics. In D. Wyse, L. Hayward, & J. Pandya (Eds.), *The SAGE Handbook of Curriculum, Pedagogy and Assessment*. London: SAGE Publications.
Hulme, M. (2008). Geographical Work at the Boundaries of Climate Change. *Transactions of the Institute of British Geographers, 33*, 5–11.
Jackson, P. (2006). Thinking Geographically. *Geography, 91*(3), 199–204.
Johnston, R. J. (1997). Australian Geography Seen Through a Glass Darkly. *Australian Geographer, 28*, 29–37.
Klafki, W. (1997). Kritisk-konstruktiv didaktik. In M. Uljens (Ed.), *Didaktik: Teori, reflektion och praktik*. Lund: Studentlitteratur.
Lambert, D. (2004). *The Power of Geography*. Retrieved February 2006, from http://www.geography.org.uk/download/NPOGPower.doc
Lambert, D., & Morgan, J. (2009). Corrupting the Curriculum? The Case of Geography. *London Review of Education, 7*(2), 147–157.
Lambert, D., & Morgan, J. (2010). *Teaching Geography 11–18 a Conceptual Approach*. Maidenhead: OUP.
Leat, D. (1998). *Thinking through Geography*. Cambridge: Chris Kington Publishing.
Lim, K. (2005). Augmenting Spatial Intelligence in the Geography Classroom. *International Research in Geographical and Environmental Education, 14*(3), 187–199.
Livingstone, D. (1992). *The Geographical Tradition: Episodes in the History of a Contested Enterprise*. London: Wiley-Blackwell.
Marsden, W. (1997). On Taking the Geography out of Geographical Education. *Geography, 82*(3), 241–252.
Massey, D. (1999). Negotiating Disciplinary Boundaries. *Current Sociology, 47*, 416–432.
Maude, A. (2016). What Might Powerful Geographical Knowledge Look Like? *Geography, 101*(1), 70–76.
Oakes, S. (2004). *World Wide Web, Geography in the News*. Royal Geographical Society. Retrieved November 2011, from http://www.rgs.org/NR/rdonlyres/B1842B47-38FD-45EB-896D-2CB407782232/0/SMA_GP_WorldWideWeb.pdf

Ofsted. (2008). *Geography in Schools—Changing Practice*. Retrieved January 2009, from www.ofsted.gov.uk

Ofsted. (2011). *Geography: Learning to Make a World of Difference*. Retrieved December 2011, from www.ofsted.gov.uk

Qualifications and Curriculum Authority. (2007). *Geography: Programme of Study: Key Stage 3*. London: HMSO.

Roberts, M. (2010). Where's the Geography? Reflections on Being an External Examiner. *Teaching Geography, 35*(3), 112–113.

Roberts, M. (2014). Powerful Knowledge and Geographical Education. *The Curriculum Journal, 25*(2), 187–209.

Rowley, C., & Lewis, L. (2003). *Thinking on the Edge*. Cumbria: Badger Press Ltd.

Royal Geographical Society Website. (2015). Retrieved December 2016, from http://www.rgs.org/GeographyToday/What+is+geography.htm

Small, J., & Witherick, W. (1995). *A Modern Dictionary of Geography* (3rd ed.). London: Arnold.

Smith, D., & Ogden, P. (1977). Reformation and Revolution in Human Geography. In R. Lee (Ed.), *Change and Tradition: Geography's New Frontiers*. London: Department of Geography, Queen Mary College, University of London.

Soja, E. (1996). *Thirdspace*. Oxford: Blackwell.

Standish, A. (2007). Geography Used to be About Maps. In R. Whelan (Ed.), *The Corruption of the Curriculum*. London: CIVITAS.

Standish, A. (2009). *Global Perspectives in the Geography Curriculum: Reviewing the Moral Case for Geography*. London: Routledge.

Taylor, L. (2009). *GTIP Think Piece: Concepts in Geography*. Retrieved February 10, 2010, from http://www.geography.org.uk/gtip/thinkpieces/concepts/#5821

Unwin, T. (1992). *The Place of Geography*. Harlow: Longman Scientific and Technical.

Vernon, E. (2016). The Structure of Knowledge: Does Theory Matter? *Geography, 101*(2), 100–104.

Walford, R. (2000). *Geography in British Schools 1850–2000*. London: Woburn.

Whelan, R. (Ed.). (2007). *The Corruption of the Curriculum*. London: CIVITAS.

White, J. (2006). *The Aims of School Education*. Retrieved February 2011, from http://eprints.ioe.ac.uk/1767/

Young, M. (2008). *Bringing Knowledge Back In: From Social Constructivism to Social Realism in the Sociology of Education*. Abingdon: Routledge.

Young, M., & Lambert, D. (2014). *Knowledge and the Future School: Curriculum and Social Justice*. London: Bloomsbury.

Young, M., & Muller, J. (2010). Three Educational Scenarios for the Future: Lessons from the Sociology of Knowledge. *European Journal of Education, 45*(1), 11–27.

4

The 'Capabilities Approach' to Geography Education

Introduction

This chapter explains the capabilities approach, the second of the main concepts at the heart of the ideas in this book. The sorts of 'powerful knowledge', embedded within a Future 3 (F3) curriculum, that was discussed in the last chapter puts the emphasis back on what the pupils get out of the educational experience, by asking in what ways their subject knowledge is powerful to them. Capabilities could provide a broader framework that links all these subjects and gives a reason for a subject-based curriculum beyond the simple instrumental notion of exam grades. There have been a number of attempts to provide this 'big picture' of a subject-based school curriculum, not least the National Curriculum 'Big Picture' (QCA 2008), the 'Trivium 21c' of Martin Robinson (2013) and the Habits of Mind (e.g. Boyes and Watts 2009) discussed previously. It is the capabilities approach which could also provide a broader picture of the school experience for young people by linking the 'powerful knowledge' they learn in the classroom to a wider set of educational aspirations.

The capabilities approach is a conceptual framework that attempts to explain what makes a 'good life', or what people judge and value to be a good life. Rather than focusing on specific measurable data such as the amount of wealth a person has, it looks at what people are capable of doing, thinking or achieving and what freedoms this affords them to live life in the way that they choose. The idea itself derives from development economics discourse and the work of Amartya Sen through a series of books and articles since the 1980s. He was awarded the Nobel Memorial Prize in Economic Sciences in 1998 for his contribution to global understanding of welfare economics. Philosopher Martha Nussbaum contributed to the understanding of the capabilities approach by refining and defining Sen's framework. Their ideas are explained in the first part of this chapter, before the ways in which educational discourse have adapted the framework are outlined, and then applied to the school subject of geography as GeoCapabilities, drawing on notions of powerful knowledge.

Development As Freedom: The Capabilities Approach to Development

The capabilities approach attempts to determine a person's capabilities, or 'capability set', which is an expression of what people are able to 'do' or to 'be'. It looks holistically at what a person is able to achieve, how they are able to think and the freedoms this affords them in life. Indeed, the title of this section, 'development as freedom', is the title of Amartya Sen's influential (1999) book on the capabilities approach. The approach is purposely a loose and fluid framework that can be applied to a wide range of peoples and societies with differing cultures, who have different ideas about what constitutes a 'good life'.

The ways in which Amartya Sen and Martha Nussbaum have developed the capability approach differs. As Robeyns (2005) explains,

> Nussbaum and Sen have different goals with their work on capabilities... Nussbaum aims to develop a partial theory of justice... as such her work on the capabilities approach is universalistic... Sen did not have such a clear objective when he started to work on the capability approach. On the one hand he was interested in the 'equality of what?' question in liberal political philosophy... on the other hand Sen was doing some much more

applied work on poverty and destitution in developing countries… (he) was also working on social choice. (pp. 103–104)

The differences can be distinguished into what Robeyns (2017) identifies as capability 'approach' and a capability 'theory'. The capability 'approach' is how Sen conceptualises the idea. It is broad, ambitious and cannot be reduced to a list of capabilities. A capability 'theory' is a much more pragmatic and practical interpretation of the notion of capabilities. As such, a list of capabilities can be drawn up for a multitude of purposes. As Robeyns (2017) argues, "there is one capability approach and there are many capability theories and keeping that distinction sharply in mind should clear up many misunderstandings in the literature" (p. 30).

It is consistent with these differences that Sen also considers 'capabilities' (as a plural) as being one capability across a number of different people whereas for Nussbaum 'capabilities' is a number of different capabilities held by the same person. For this reason, the idea of a 'capability set' is used to express a collection of different listable (but not necessarily measurable) capabilities held by a particular person.

Despite these differences, the literature on the capability approach includes a range of terminology as part of a broad conceptual framework. A person's capability is the product of their 'commodities'. This includes a person's wealth, the health of their diet, the standard of their living conditions and the level of their education. These commodities can go through a series of 'conversion factors' to create a 'capability set', a collection of knowledge, skills and competencies. The creation of the capability set can be inhibited by 'capabilities deprivation', any factor that inhibits the gaining of capabilities. The larger the capability set the more 'functioning' this can afford a person, which are the 'beings and doings' of everyday life. Choice about how to function is a key aspect of the framework. The purpose of the functioning is 'utility', what a person wants to achieve in the world to make them happy and fulfilled. Sen makes a further distinction with 'agency', which is about living and thinking critically as part of a society, with practical considerations such as the taking on of a political cause which may not lead to happiness but are considered an important aspect of a 'free' adult life. These aspects of the capabilities approach can be illustrated in a simple model, shown in Fig. 4.1.

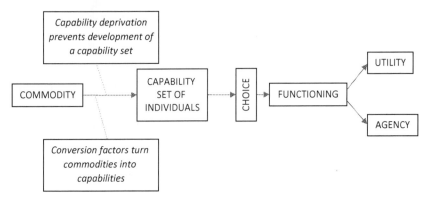

Fig. 4.1 A conceptual model to show the various aspects of the capabilities approach

The capabilities approach can be illustrated with reference to development discourse. With more income, people are able to afford more food and better accommodation, all of which counts as the 'commodity' input. This in turn could enable them to be fitter and healthier people who thus have more capabilities, and a better developed capability set to live in the world in the way they would want to. Though we must be careful, as Robeyns (2003) warns,

> commodities … should not necessarily be thought of as exchangeable for income or money—as this would restrict the capabilities approach to analyses and measurement in market-based economies, which it does not intend. (p. 12)

As she warns, commodities as an input to the capability approach are not just about money; she includes aspects of market (and non-market) production, income and other forms of 'transfers in kind'. Rather than these leading directly onto capability, commodities need to undergo three 'conversion factors'—*personal* factors which include metabolism, physical condition and intelligence; *social* factors which include public policies, social norms and power relations; and *environmental* factors, which include climate, infrastructure, institutions and public goods. To illustrate the relationship between commodities

4 The 'Capabilities Approach' to Geography Education

and capability she uses an analogy of a bicycle as a means to achieve mobility. The bicycle is the commodity input and the mobility is a 'capability'. Yet as she argues,

> if a person is disabled, or in a bad physical condition, or has never learned to cycle, then the bike will be of limited help to enable the functioning of mobility... If there are no paved roads, or if a society imposes a social or legal norm that women are not allowed to cycle without being accompanied by a male family member, then it becomes much more difficult or even impossible to use the good to enable the functioning. (Robeyns 2003, pp. 12–13)

Sen (1985b, 1999) does not identify any specific capabilities that may be part of a capability set, as he argues that capabilities will vary across societies, though some have argued (e.g. Nussbaum 2000 and Finnis 1980) that 'bodily health' would be a key universal capability as without this, the ability to carry out any sort of functioning in life would be restricted. A key contention in the capability literature is the extent to which a 'list' of universal and valuable capabilities can be defined. Sen (1985a, b, 1999) has consistently refused to define his own, or to endorse others' lists of capabilities. For Sen, the capabilities approach remains a philosophical framework and lists of capabilities will be specific to each society or group of people wanting to create them. This gives the capabilities approach the dynamism to be interpreted and evaluated in many different ways by different societies. As soon as a list is created or endorsed it immediately becomes reduced to a check list, and this tick box approach to defining a 'good life' removes the complexity of the approach. Despite this, Clark and Qizilbash (2002) suggest Sen often provides examples of intrinsically valuable capabilities such as being able to live long, escape avoidable morbidity, be well nourished, be able to read, write and communicate, and take part in literary and scientific pursuits. This lack of a specific list of capabilities has led others to offer their own sets of universal capabilities in an attempt to further Sen's framework (e.g. Alkire and Black (1997), Alkire (2002), Clark and Qizilbash (2002) Clark (2003), Nussbaum (1990, 1995, 2000, 2003) and Robeyns (2003)). Table 4.1 offers an overview of selected capability sets.

Table 4.1 A collection of selected 'universal capabilities', based on Alkire (2002)

Finnis (1980)	Griffin (1986)	Galtung (1980)
Life	Accomplishment	Input-output (nutrition, water, air)Climate balance with nature (clothing, shelter)
Survival	Components of human existence	Health
Health	Deciding for oneself/agency	Community
Reproduction	Minimum material goods	Symbolic interaction and reflection (education)
Knowledge	Limbs & senses that work	
Meaningful	Freedom from pain & anxiety	
Work/livelihood	Liberty	
Authentic self-direction	Understanding	
Participation/agency	Enjoyment	
Relationships	Deep personal relations	
Inner peace		
Environment & aesthetic	**Lasswell and Holmberg (1969)**	**Qizilbash (1996)**
Davitt (1968)	Skill	Health/nutrition/sanitation/rest/shelter/security
Life and reproductionProtection and securityTitle (property)	Affection	Literacy/basic intellectual and physical capacitiesSelf-respect and aspiration
Sexual union	Respect	Positive freedom, autonomy or self-determination
Decision-responsibilityKnowledge	Rectitude	Enjoyment
Art	Power	Understanding or knowledge
Communication	Enlightenment	Significant relations with others and some participation in social life
Meaning	Wealth	Accomplishment
	Well-Being	

4 The 'Capabilities Approach' to Geography Education

These authors have all adopted different interpretations of capabilities and used different methodologies and ideas to create their lists, hence the different ideas proposed. There are many similarities within these sets, with notions of 'health' and 'life' appearing on multiple lists. However, it is Nussbaum's (2000) list of capabilities that has become the most celebrated in academic literature and is listed below:

1. *Life.* Being able to live to the end of a human life of normal length; not dying prematurely, or before one's life is so reduced as to be not worth living.
2. *Bodily Health.* Being able to have good health, including reproductive health; to be adequately nourished; to have adequate shelter.
3. *Bodily Integrity.* Being able to move freely from place to place; to be secure against violent assault, including sexual assault and domestic violence; having opportunities for sexual satisfaction and for choice in matters of reproduction.
4. *Senses, Imagination and Thought.* Being able to use the senses, to imagine, think, and reason—and to do these things in a "truly human" way, a way informed and cultivated by an adequate education, including, but by no means limited to, literacy and basic mathematical and scientific training. Being able to use imagination and thought in connection with experiencing and producing works and events of one's own choice, religious, literary, musical and so forth. Being able to use one's mind in ways protected by guarantees of freedom of expression with respect to both political and artistic speech, and freedom of religious exercise. Being able to have pleasurable experiences and to avoid non-beneficial pain.
5. *Emotions.* Being able to have attachments to things and people outside ourselves; to love those who love and care for us, to grieve at their absence; in general, to love, to grieve, to experience longing, gratitude, and justified anger. Not having one's emotional development blighted by fear and anxiety. (Supporting this capability means supporting forms of human association that can be shown to be crucial in their development.)
6. *Practical Reason.* Being able to form a conception of the good and to engage in critical reflection about the planning of one's life. (This entails protection for the liberty of conscience and religious observance.)

7. *Affiliation.*

- Being able to live with and toward others, to recognise and show concern for other humans, to engage in various forms of social interaction; to be able to imagine the situation of another. (Protecting this capability means protecting institutions that constitute and nourish such forms of affiliation, and also protecting the freedom of assembly and political speech.)
- Having the social bases of self-respect and non-humiliation; being able to be treated as a dignified being whose worth is equal to that of others. This entails provisions of non-discrimination on the basis of race, sex, sexual orientation, ethnicity, caste, religion, national origin and species.

8. *Other Species.* Being able to live with concern for and in relation to animals, plants, and the world of nature
9. *Play.* Being able to laugh, to play, to enjoy recreational activities.
10. *Control over One's Environment.*

  (a) *Political*: being able to participate effectively in political choices that govern one's life; having the rights of political participation, free speech and freedom of association.
  (b) *Material*: being able to hold property (both land and movable goods); having the right to seek employment on an equal basis with others. (Nussbaum 2000).

As she argues, this list "isolates those human capabilities that can be convincingly argued to be of central importance in any human life, whatever else the person pursues or chooses" (Nussbaum, 2000, p. 74). Her ambition was to create a list of universal capabilities. Yet her list has been criticised as offering a white, North American, middle class viewpoint of what everyone in the world ought to crave. Stewart (2001) complains that Nussbaum did not listen to the voices of the poor when generating her list. Clark and Qizilbash (2002) are even more critical, arguing:

[O]n closer inspection, however, Nussbaum's...theory...turns out to be derived largely from Ancient Greek Philosophy instead of concrete studies of human values. (p. 3)

Their criticism stems from a seeming lack of empiricism. Yet similar criticisms could be made of each of the capability sets offered in Table 4.1, which were created by different writers for varied reasons.

Sen's refusal to endorse a set list of specific capabilities has given rise to debate between Nussbaum and Sen. As Nussbaum (1988) argues, Sen

needs to be more radical than he has been so far ...by describing a procedure of objective evaluation by which functionings can be assessed for their contribution to the good human life. (p. 176)

She accepts he will not endorse a list, but wants him to at least identify the means by which capabilities can be decided upon. Sen (2004) has been equally critical of Nussbaum's (2000) list, arguing,

[T]o have such a fixed list, emanating entirely from pure theory, is to deny the possibility of fruitful public participation on what should be included and why. (Sen 2004, p. 77)

For Sen, it is societies who decide on their own capabilities through participation and should not have any list imposed on them. In an attempt to reconcile these two very different viewpoints, Gasper (2004) argues that Nussbaum's list should be a starting point for discussion "in each society, as a rational interpretation, implication and evolution of their values" (Gasper 2004, p. 187). Discussions between individuals in communities can enable capabilities to evolve, based on what that society values, and the capabilities approach provides a structured framework for those discussions to take place.

The discussions outlined here in the development discourse illustrate a key antagonism inherent within the capabilities approach. On the one hand, it is a conceptual, philosophical framework that presents an idea of development and a means to discuss what is valued in life. On the other hand, there is a need to define capabilities in more practical terms, to create a capability 'theory' from the broader 'approach', which would render it much more applicable and useful in real-life contexts.

The creation of 'capability' in people can lead on to a variety of functionings. The 'functioning' is a person actually carrying out the doings and beings they are capable of, such as going to work to earn an income, eating a healthy diet, or living in a safe and secure home. Again, Sen does not offer any specific functioning that is of value as again this will vary in time and space, and will be dependent on the capability set of the individual.

Yet for a number of writers the capabilities themselves are the valuable end point of the theoretical framework. As Robeyns (2017) argues:

> In developing a capability theory, we need to decide whether we think that what matters are capabilities, functionings, or a combination of both. (p. 66)

She gives a practical example, arguing,

> being knowledgeable and educated can very plausibly be seen as of ultimate value, but it is of instrumental value for various other capabilities… being healthy, being able to pursue projects, being able to hold a job and so forth. (p. 55)

Capabilities and functioning are two similar ideas, but are significantly different. A person with a capability of bodily health will be able to function effectively by choosing to go out to work to earn an income. The capability is bodily health; the functioning is going to work. As Sen (1987) explains,

> a functioning is an achievement, whereas a capability is the ability to achieve … Capabilities… are notions of freedom, in the positive sense: what real opportunities you have regarding the life you may lead. (p. 36)

A person with the capability of bodily health uses that to go to work and earn money (the functioning), this then gives them financial security, fulfilment and happiness (the utility). The idea of 'utility' will also differ across the globe; what people consider to be valuable outcomes in life by functioning in particular ways will be affected by cultural and social factors as well as societal norms and historical considerations. What is distinctive in the capabilities approach is that people are able to achieve the utility that they crave, such as happiness or emotional security, and this enables true freedom and thus a more complete picture of 'wellbeing'.

The key to the framework is the personal freedom and *choice* people have about how to behave with the given capability set, hence the idea of 'development as freedom' (Sen 1999). For Sen (1999) measuring a country's development through statistics such as GDP says nothing about the ability of the people in those countries to live a full and fulfilling life in which they are able to make choices about how to live. According to Sen (1999), this is what the capabilities approach enables, an understanding of the real freedoms that people have. This says far more about development than traditional statistics.

A further aspect of the capabilities approach is the identification of 'capability deprivation', which describes any factor that inhibits the development of specific capabilities. These 'negative freedoms' as Sen would identify, "stem from the violation of personal rights as well as the absence of positive freedoms" (Clark 2005 p. 9). As Iceland (2004) defines, "poverty should be viewed as the deprivation of basic capabilities rather than merely the lowness of income" (p 1). In development discourse, factors that inhibit the development of individuals can be regarded as forms of capability deprivation and might include, for example, rules regarding the role of women in society which would restrict girls from going to school. These rules inhibit the gaining of capabilities which an education would provide, so these rules are seen as 'capability deprivation'.

The capability approach and the various theories it has inspired have had an impact across a range of disciplines. As Robeyns argued in 2003,

> the next decades will show whether the capabilities approach remains primarily a philosophical framework, or whether it will grow into a mature paradigm for well-being, development, and social policy. (p. 54)

For her, the capabilities approach has much promise and the approach has since been taken on in other academic fields, including in education. As Hinchliffe and Terzi (2009) explain, "the time for capabilities for educational researchers, writers and thinkers seems to have finally arrived" (p 387), and the thinking led to the notion of GeoCapabilities and the ideas in this book.

Education As Freedom: The Capabilities Approach to Education

The capabilities approach to educational thinking can offer a means of expressing what freedoms an education allows a person to 'be' or to 'do'. Rather than discussing the success of education based on figures such as pass rates, or exam grades, a person's education enables them to develop a set of capabilities to allow them to function in the world. Thus, the capabilities approach asks what an education can enable a young person to achieve far beyond any instrumental measure of success. In short, it facilitates a focus on the 'outcomes' of education rather than simply the measurable 'outputs'.

Much of the development of the thinking on capabilities in education has come from Martha Nussbaum herself (2006, 2016) in her work on democratic citizenship. For her, democratic citizenship is a key means to achieve the list of 10 universal capabilities outlined earlier. Intelligent people are able to use their democratic rights to elect officials who will be able to enact the universal capabilities she identified. Yet full democratic citizenship is threatened by what she identifies as a 'silent crisis' in education, which is narrowly focussed on numeracy and literacy, and driven by the demands of industry. As she argues:

> Radical changes are occurring in what democratic societies teach the young, and these changes have not been well thought through. Thirsty for national profit, nations, and their systems of education, are heedlessly discarding skills that are needed to keep democracies alive. (Nussbaum 2016, p. 2)

In response, she argues critical thinking, citizenship of the world and narrative imagination are three important tenets of a school curriculum that can create a population capable of democratic participation. The second of these has obvious connections to the ideas of a geography education, as she has also argued:

> [T]he abilities connected with the 'humanities' and the 'arts' are crucial to the formation of citizenship. They must be cultivated if democracies are to survive, through educational policies that focus on pedagogy at least as much as on content. (Nussbaum 2006, pp. 387–388)

Nussbaum has recognised the significance of a humanities education as being connected with a broader set of educational goals. She has articulated these as being able to participate in democracy, what could be seen as a 'functioning', with the knowledge gained in education as a 'capability'. Yet she has defined humanities and arts as skills and abilities to cultivate, rather than seeing any value in the knowledge of these subjects, and her solution is to focus on pedagogy. This thinking runs the risk of creating a Future 2 (F2) curriculum to address her concerns, without fully engaging with the knowledge content of subjects.

Despite this critique, she started the important work which links subjects with broader educational capabilities. By the 2010s, there had been many attempts to introduce the capabilities approach to areas of educational study (all discussed in Walker 2006): Page (2004) and Raynor (2004) consider capabilities in teacher development; Terzi (2003, 2004) considers capabilities in special education; capabilities and school leadership are discussed in Bates (2004); capabilities and adult literacy are discussed in Alkire (2002); and capabilities and higher education, the largest field of capabilities education research, is discussed in Flores-Crespo (2001), Deprez and Butler (2001), Watts and Bridges (2003), Bozalek (2004) and Walker (2006). When discussing the nature of research into the capabilities approach to education, Hinchliffe and Terzi (2009) identify two areas of research:

> One focuses on the nature of capabilities themselves, how they are to be developed and what kinds of functioning their development is likely to afford. The second approach focuses more on the structural features (institutional, social, economic) that govern the development of capabilities. (p. 388)

These 'structural features' determine the nature of a capabilities set that develops, with factors limiting capabilities development again being recognised as 'capability deprivation'. The capability set then enables a person to have choices in how to function in the world, including the taking on of agency. In education, Hinchliffe (2006) has identified a further type of functioning, 'occupational functioning', to express choices a person would be able to make about their career or job. With a larger capabilities set, the choice of career would be much wider. Figure 4.2 is a model to identify the way the capabilities approach to education has been discussed in the literature.

In Fig. 4.2, the concept of 'structural features' derives from work on 'structuration theory' (e.g. Giddens 1984). This work suggests that people's actions are the product of a set of societal norms and values which are often unwritten. These norms provide a 'structure' which informs and affects the decisions people take. Structures are often seen as being 'constraining', affecting the ability of an individual to exercise complete free will. In education, this would identify that teachers operate within a set of real and perceived professional 'structures': teachers have to be based predominantly in a school classroom, have to teach within a set amount of time, and have to follow a curriculum structure, for example. Teachers have to work within these structures at a variety of scales, and 'structural features' is a way of expressing the nature of these constraints. Yet Giddens (1984) "redefines the role of structure by realising that it can be both a constraining and enabling element for human action" (Lamsal 2012, pp. 112–113), suggesting that people are able to react to and interact

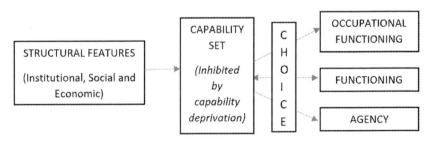

Fig. 4.2 A conceptual model of how the capabilities approach has been applied to educational discourse

4 The 'Capabilities Approach' to Geography Education

with potential structures. These structures can be constraining to teachers' work, and can therefore be a negative influence or can be 'enabling', providing more opportunities for teachers. This idea of 'structuration' is an attempt to conceptualise this relationship between the structures imposed on an individual and their response to it.

Structural features in education are identified by Hinchliffe and Terzi (2009) in terms of institutional features, social features and economic features. Institutional features are concerned with the place in which the education is taking place. It refers to an institution's policies and ethos as well as national policies of compulsory education provision. It also includes aspects of the curriculum, how learning is organised, examination criteria, the nature of teaching and learning, and resources. 'Social' features are concerned with the nature of the learner, their capacity to learn, social background, income and age. 'Economic' features include the traditional elements of 'commodity' (from the original conception of capabilities), how much finance there is in education and what this can afford the institution in terms of resources, learning environment, and extracurricular opportunities, which may also impact on student capabilities. These structural feature elements need to combine, as Hinchliffe and Terzi (2009) explain:

> A wonderful pedagogy with first rate teachers will have little impact without being embedded in the right kind of institutional framework. On the other hand, excellent policy initiatives that promote favourable institutions and resources will have little impact (so far as capabilities is concerned) if the curriculum is impoverished and does not address the development of capabilities. (p. 388)

For them, the institutional framework of schools affects the nature of the curriculum and this promotes the development of capabilities in pupils. Structural features can also inhibit the development of capabilities, as Saito (2003) argues,

> if the education system takes an extremely 'top down' approach and stresses competitiveness, children tend to study subjects that are required for examination success... In this case, the children ... are considered to have limited capabilities. (p. 27)

This top down competitive nature of curriculum, a feature of the contemporary education system dominated by league tables, could be seen as a form of capability deprivation as the idea of freedoms and choice as a result of education is removed in the quest for examination grades.

The pupil develops a capability set as an outcome of their education and this enables them to make choices about how to live. As White (1973) argues, "in order for the child to be able to make choices in his/her life, the child needs to become autonomous through education" (cited in Saito 2003, p. 27). An educated person is free to make decisions for themselves, choosing their functioning, as modelled in Fig. 4.2. The more developed the capabilities set, the more choices are available to that educated person. Functioning takes many forms; it is the beings and doings of life, 'occupational functioning' about what career to follow and the taking on of 'agency', being active and critically engaged in the twenty-first century society. Agency is about the capacity of individuals to make choices, exercising free will to think and behave in their own 'free' way. Agency can include the taking on of a political or environmental cause by joining a pressure group or campaigning. However, if teachers actively promote particular non-public values to students, perhaps as part of a hidden curriculum, this could be seen as a form of 'capability deprivation'.

Prior to discussions of 'GeoCapabilities', the literature on the capabilities approach to education had two major omissions and these relate to the role teachers play in enhancing the capabilities of pupils and the place of subject knowledge in developing such capabilities. Both of these, however, are implicit the literature. Hinchliffe and Terzi (2009) identify 'first rate teachers' who are able to help pupils to develop capabilities although they never expand on exactly what they mean by this; the implication in their writing is that this refers to the pedagogical ability of teachers. The importance of a subject-based curriculum to develop capabilities is less clear. Saito (2003) assumes a natural role for subjects, as he illustrates:

> Lisa learns mathematics and as a result, she has … newly created opportunities and capabilities… (which) may be ones that Lisa was not aware of, and which were not in her capability set before learning mathematics.

For Saito (2003), there is a direct link between subject knowledge and capabilities, which supports the arguments for a subject-based curriculum. However, what is less clear in the literature on educational capabilities is the process that links teachers' work, subjects and capabilities. This gap in the research and the possibilities of the idea of capabilities and the Future 3 curriculum providing a response to this gap underpinned the research into GeoCapabilities and has led to the ideas in this book.

However, other writers disagree about the importance of subjects developing capabilities. In order to develop capabilities to enable 'deliberation over career choice', Hinchliffe (2009) argues,

> space in the curriculum needs to be made for career choice, from school right through to university" (p. 412)... so much time and attention is devoted to the acquisition of knowledge, understanding and skills whilst crucial decisions consequent on this education (what university shall I apply for, what course should I pursue, which occupation best suits me) are often left... to chance. (p. 412)

For Hinchliffe (2009), capabilities which enable occupational functionings require their own curriculum time and cannot be developed through traditional school subjects. This undermines the role of a subject-based curriculum. These discussions can be related back to the three curriculum futures heuristic developed by Young and Muller (2010). If a curriculum was designed specifically to develop 'occupational functions' this could create an F2 curriculum, through a 'vocational' ideology. In this scenario, even if a school had a subject-based curriculum the careers education would be entirely separate; in fact, subject knowledge would be a distraction. Often very crude links are made between certain subjects and certain careers, such as the study of English to become a lawyer, for example. Some subject knowledge is required directly for certain careers, with careers in medicine requiring knowledge of chemistry being the obvious example, but these specific requirements are only applicable for a small range of careers. In the popular imagination of the subject, geography students are often told they can become meteorologists or geography teachers, which is often unhelpful advice for potential students.

Geographers are amongst the most employable graduates (RGS-IBG 2016) given the broad range of knowledge the students develop. What the capabilities approach to this debate suggests is that far from being a distraction, and far from a crude link, powerful subject knowledge can enable choice of careers. It is subject knowledge that enables young people to be able to engage in debate and discussion, to think in new ways, to be able to discern fact from fiction, and these qualities develop capabilities that are of value not only in the adult world but in any workplace. At a practical level, pupils are introduced to a wide range of careers through the study of subjects. In geography, pupils can learn about the role and work of urban planners, climatologists, lawyers, aid workers and government officials amongst a whole range of professions. Schools do have a responsibility to teach young people about possible career choices and may well take time out of the curriculum to run work experience days, specific practical advice and workshops. But an F3 curriculum, embedded with powerful knowledge, sees subject knowledge playing a central and significant role in the preparation of young people for their work futures.

Educational Capabilities Lists

When applied to development discourse, Sen always avoided listing specific capabilities for students to aspire to as part of the capabilities approach, and so for education Saito (2003) asks, "[I]s it possible to outline a range of capabilities that children should engage with?" (p. 29). A list of capabilities could form the basis of curriculum planning in schools, but equally could be seen as restrictive and reduce curriculum to a tick list of competencies. The closest attempts to define lists of universal educational capabilities come from the fields of special educational needs (Terzi 2005) and higher education (Walker 2006). Their lists (slightly abridged) are shown in Table 4.2.

These lists were created for different reasons and for different types of learners. Terzi's (2005) list is for learners with disabilities and those with special educational needs, and the "capabilities approach helps… to (define) fundamental educational capabilities at levels necessary to

Table 4.2 A list of suggested universal 'educational' capabilities

Terzi (2005)	Walker (2006)
Literacy: Being able to read and to write, to use language and discursive reasoning functionings. **Numeracy:** Being able to count, to measure, to solve mathematical questions and to use logical reasoning functionings. **Sociality and participation:** Being able to establish positive relationships with others and to participate without shame. **Learning dispositions:** Being able to concentrate, to pursue interests, to accomplish tasks, to enquire. **Physical activities:** Being able to exercise and being able to engage in sports activities. **Science and technology:** Being able to understand natural phenomena, being knowledgeable on technology and being able to use technological tools. **Practical reason:** Being able to relate means and ends and being able to critically reflect on one's and others' actions.	**Practical Reason:** Being able to make well-reasoned choices. **Emotional resilience:** Able to navigate study, work and life. **Knowledge and imagination:** Being able to gain knowledge of a chosen subject—disciplinary and/ or professional—its form of academic enquiry and standards. Being able to use critical thinking and imagination to comprehend the perspectives of multiple others and to form impartial judgements. **Learning dispositions:** Being able to have curiosity and a desire for learning. **Social relations and social networks:** Being able to participate in a group for learning, working with others to solve problems and tasks. **Respect, dignity and recognition:** Being able to have respect for oneself and for and from others. **Emotional integrity and emotions:** Not being subject to anxiety or fear which diminishes learning. **Bodily integrity:** Safety and freedom from all forms of physical and verbal harassment.

function and participate effectively in society" (p. 7). For Walker (2006), working in higher education, "the idea is for higher education communities… to produce their own flexible, revisable and general list" (p. 49). What is common in both these lists is the idea of the capabilities approach expressing the ideal conditions to learn, rather than any specificities of a curriculum. The focus is on the individual structural features of curriculum, such as the personal qualities of a learner, with the idea of 'learning dispositions' and 'practical reason' being a feature of both lists. These lists can provide a starting point for teachers to think about the pupil focussed outcomes of a school curriculum, but a direct curricular translation is evocative of F2 curriculum thinking.

Both lists support the development of capabilities through knowledge. For Terzi (2005), knowledge is in the form of literacy, numeracy, science and technology whereas Walker (2006) identifies "knowledge of a chosen subject—disciplinary and/ or professional—its form of academic enquiry and standards" as a capability in itself. This 'form of enquiry and standards' of a subject identifies that a subject is more than simply a list of facts but is a way of 'thinking' about knowledge, a sentiment which is close to the ideas of powerful knowledge and F3 curriculum thinking. The values dimension of knowledge is also an explicit consideration of capabilities. What Walker (2006) is advocating is 'open mindedness', and an 'awareness' of ethical debates and moral issues, as well as 'listening to and considering other person's points of view in dialogue and debate'. It is an example of values clarification, where complex issues are clarified and this then allows a student to decide their own position on the issue rather than being told what to think as a result of their education, or values 'transmission' from teacher/lecturer to student. Being told what to think, rather than how to think, would be an example of capabilities deprivation and the capabilities approach is able to articulate this difference.

These debates can be of value to secondary school teachers when 'curriculum making'. By ensuring subject knowledge is explicitly defined as a type of capability, it advocates a subject-based curriculum. Specific subject-based capabilities lists have started to emerge in academic literature. Sharp and Watts (2004) explore the role that capabilities might play in students of religious education (RE) by interviewing former RE students and assessing the extent to which they integrate notions of their RE education in their everyday lives. Whilst they discuss 'capabilities', they do not explore the concept in detail nor do they offer a list of what RE capabilities might be. Hinchliffe (2006) has devised a list of capabilities for humanities students in higher education (Table 4.3).

The creation of this list suggests that humanities education is about more than simply acquiring knowledge for the sake of passing a humanities exam. The capabilities approach tries to determine what that knowledge can enable a person to be like, to do and to think and that provides a justification for the study of that subject. Through a humanities education, a person can develop the capabilities to have, for example, 'practical judgement'. This is expressed directly as an outcome of a humanities

Table 4.3 A list of capabilities derived from the study of humanities in higher education (Hinchliffe 2006)

Hinchliffe (2006)
1. Critical examination and judgement.
2. Narrative imagination.
3. Recognition/concern for others (citizenship in a globalised world).
4. Reflective learning (ability to articulate and revise personal aims).
5. Practical judgement (in relatively complex situations).
6. Take responsibility for others.

education. Teachers could use the humanities list of capabilities as a starting point for curriculum design, to then plan learning activities to help young people engage directly with the knowledge that will enable those capabilities to develop.

Values also play a role in humanities capabilities. The list encourages students to develop 'recognition/concern for others'. Teachers would need to be careful in interpreting these capabilities for use with students to ensure any 'concern' they were discussing promoted 'public values' and not the teachers' opinions. 'Concern' could be misinterpreted as encouraging 'agency' whereby values are transmitted, rather than enabling the 'choice' element of the capabilities approach.

The problem with the list of humanities capabilities, however, is that it is not unique to humanities. Other subjects could cite 'reflective learning' as a desirable capability for students to develop. If all these capabilities could be developed through other subjects on a school curriculum, then in an overcrowded curriculum there would be no need for humanities to be part of a curriculum at all. This list also fails to engage with the knowledge content of humanities, expressing the subject through competencies, which follows F2 curriculum thinking. However, if a list of capabilities were created that is unique to a particular subject, and the only way to develop those capabilities was through studying that particular subject, then the capabilities approach could be a means to express the value of that subject in a curriculum. This is where the notion of powerful knowledge as discussed in the previous chapter can be of use; the powerful knowledge of subjects can enable capabilities to develop. It is this contention, argued through the subject of geography, that underlies the notion of GeoCapabilities.

The Capabilities Approach to Geography Education: GeoCapabilities

GeoCapabilities is the culmination of two broad sets of ideas discussed in this book. Firstly, the powerful knowledge of geography provides the 'geo', and secondly the capabilities approach of this chapter provides the 'capabilities'. It is a conceptual framework which offers a way of thinking about the contribution that geographical knowledge makes to an educated person. It is an expression of how powerful geographical knowledge can enable children to think and behave in ways that promote freedoms in life. It articulates what studying geography attempts to achieve that no other subject can. It is broad and ambitious, and attempts to provide a fill for the conceptual gap in the existing ideas by linking teacher work, knowledge and capabilities. Lambert and Morgan (2010) first introduced GeoCapabilities, with the idea being developed in Lambert (2011a, b, 2016) and then in two articles developed as part of the first two GeoCapabilities projects (Solem et al. 2013, Lambert et al. 2015).

A key contention within discussions around the capabilities approach has been the extent to which 'lists' of capabilities can be drawn up. GeoCapabilities has been subject to a similar discussion. When Lambert (2011b) initially presented GeoCapabilities, he resisted the temptation to define a distinct 'list' of GeoCapabilities, in a similar vein to Sen's thinking (1985a, b) in initial conceptions of the capabilities approach. By not defining GeoCapabilities, the concept remains a theoretical framework and teachers themselves are in a position to decide what they consider to be important outcomes of geography education, related to their ideologies of the subject and the way it fits within broader ideas of education. As soon as a list of GeoCapabilities is produced, it might reduce the concept to a tick list, which would restrict curriculum thinking.

Despite not clarifying what GeoCapabilities might be, Solem et al. (2013) define and list "three GeoCapabilities" shown in Table 4.4.

This list of GeoCapabilities is in fact a reworking of three of Nussbaum's ideas about universal capabilities, "phrased in a manner that enables analysis of the curricular role of geography" (Solem et al. 2013, p. 216). This list, however, much like the list of humanities capabilities from Hinchliffe

4 The 'Capabilities Approach' to Geography Education

Table 4.4 A suggested list of the capabilities approach to geography: 'Three GeoCapabilities' from Solem et al. (2013, p. 221)

Solem et al. (2013)
1. Promoting individual autonomy and freedom and the ability to use one's imagination and to be able to think and reason.
2. Identifying and exercising one's choices in how to live based on worthwhile distinctions with regard to citizenship and sustainability.
3. Understanding one's potential as creative and productive citizens in the context of the global economy and culture (p. 221).

(2006), does not adequately articulate the knowledge component of geography or the unique nature of the subject. Other subjects could assist in the development of these capabilities, most significantly the subject of citizenship. Whilst these could be considered a list of educational capabilities based on Nussbaum (2000), it does not express 'geo' capabilities.

To bring the 'geo' to the forefront of discussions about the capabilities approach to geography education we can again return to the discussion of powerful knowledge from the last chapter. It is the notion that what makes knowledge powerful is determined by how it is to be considered and understood in the minds of the learners, which links powerful knowledge to capabilities. GeoCapabilities are defined by the powerful geographical knowledge on which they are based and what thinking this can enable in a young person. This notion was first considered by Lambert and Morgan (2010); the capabilities approach provides powerful knowledge with a rationale. Based on this idea, Lambert and Morgan (2010) identified three 'expressions of powerful knowledge' with which GeoCapabilities can be developed. Successive publications have since developed the concept (Lambert 2011a, b, 2016), with Maude's types of knowledge discussed in the last chapter being incorporated into Lambert's (2017) version, shown in Table 4.5.

These 'expressions' of powerful knowledge do not simply list facts to be learnt. They are broader than Maude's (2016) typology of powerful knowledge and refer specifically to geographical knowledge. 'Deep descriptive world knowledge' does not simply mean lists of capital cities or place names, which could be indicative of an F1 (Future 1) curriculum. The idea goes beyond this and is about developing a sense of how places come to be; this could be through a positivist scientific investigation into

Table 4.5 Expressions of the powerful knowledge of geography on which GeoCapabilities is based, with reference to Maude's types of geographical knowledge (2016) (Lambert 2017)

Expressions of powerful geographical knowledge on which capabilities depend
• *The acquisition of* deep descriptive and explanatory 'world knowledge'. This includes, for example, countries, capitals, rivers and mountains, also world wind patterns, distribution of population and energy sources. The precise constituents and range of this substantive knowledge are delineated locally influenced by national and regional cultural contexts. **TYPE 2**
• *The development of* the **relational thinking** that underpins geographical thought. This includes place and space (and scale), plus environment and interdependence. This knowledge component is derived from the discipline. Thus, these 'meta-concepts' are complex, evolving and contested. **TYPE 1 TYPE 3**
• *A propensity to* apply the analysis of alternative social, economic and environmental **futures** to particular place contexts. This requires appropriate pedagogic approaches such as decision-making exercises. In addition to intellectual skills such as analysis and evaluation this also encourages speculation, imagination and argument. **TYPE 4**

a place or understanding people's emotional response to place. This knowledge is 'deep' in the sense that it requires detail, backed up by evidence, but 'substantive' to express the idea that the potential content could be vast in scale. The words "and explanatory" were added in Lambert (2016) after 'deep descriptive' to clarify the idea that understanding how places are and how they come to be is an important consideration, incorporating Maude's (2016) ideas of 'analysis' and 'understanding'. Children themselves could have a hand in developing this knowledge through their own fieldwork investigations but equally can engage with distant places through other means. Teachers reveal the world to children through the ways they frame and recontextualise knowledge to develop their world knowledge. No other subject offers this. Other subject teachers might mention places in the world, and they may even take the time to show children where they are on a world map but this is superficial. Those teachers will be simply using the place to illustrate a phenomenon from their own subject. However, deep descriptive world knowledge treats places as unique, individual and worthy of deep thought and is what occurs in good geography lessons.

4 The 'Capabilities Approach' to Geography Education

The second expression, about 'relational understanding' of people and places in the world, is an articulation of physical and human processes. These are theoretically informed; the understanding of the world is derived from a set of thought processes that are distinct to the subject of geography. These create processes such as 'migration' or 'erosion'. The concepts themselves may not be bounded by a specific place but can be applicable to a variety of places. The idea of 'relational thinking' (Lambert 2016) expresses how people are related to other people and how places are related to other places through 'thinking geographically' about the world (Jackson 2006). This could be through broad physical and human processes, but it also expresses how people and place are interrelated, such as in the understanding of climate change as articulated by Hulme (2008). Central to 'relational understanding' is how young people relate to their geographical knowledge and this is why Maude's (2016) type 1 and 3 knowledge are about children taking power over their own knowledge to develop understanding of their place in the world.

The third expression of powerful knowledge on which GeoCapabilities is based suggests that geography can enable children to think about alternative social, economic and environmental 'futures'. This has a number of considerations. Firstly it is futures orientated, forcing children to envision life that has not yet taken place. It encourages them to think about how they can contribute to shaping their future, and this links clearly with developing agency, the ideas of a moral education and developing a sense of responsibility for themselves and the world. This is engrained into the way students think about the world, hence the 'propensity' to think about these futures. The 'application of analysis' requires the specific thinking of the geographer developed through the discipline of geography. A variety of futures is also a key consideration, how the world might change socially, economically and environmentally, and this offers a variety of viewpoints. It is this understanding that enables Maude's (2016) aspiration for young people to be able to participate in debates and discussions over their futures. However, it also encourages children to think about physical and human processes beyond their control such as cycles of erosion and deposition, climatic changes and how the world might change over both human and geological timescales.

The expressions of powerful geographical knowledge here also enable the vignettes of powerful knowledge that were illustrated in the last chapter to be placed within this framework. The vignette of coastal geomorphology (Table 3.3) can thus be re-examined. When pupils learn about the Holderness coastline they are engaging with all three of these expressions of powerful knowledge; they learn the location of the Holderness coast alongside the North Sea and its latitude on the earth; this understanding enables this relational thinking to develop—the latitude in turn affects weather patterns, which creates the erosive power of the sea. The relationship between geology, erosion, weather and climate develops the understanding of the geography of this place; pupils also gain an understanding of the future of the coastline in terms of longer term changes to climate. It is this more holistic understanding of geographical knowledge that in turn can enable a young person to develop capabilities to think about the world in new ways, thinking geographically about sustainability, cycles of erosion and climate change. It is this broader understanding that contributes to young people's capabilities to make informed choices about how to live and think in the modern world.

Lambert (2016) describes the powerful knowledge of geography discussed here as a 'bridge' between the aims and aspirations of a geography curriculum and the development of GeoCapabilities in young people. Figure 4.3 models this relationship, where powerful knowledge as expressed by Lambert and Morgan (2010) is the bridge between the curriculum and the outcomes of capabilities as defined by Solem et al. (2013). Curriculum making by teachers is the process by which this can occur, and as such teachers play a fundamental role in the development of GeoCapabilities in their pupils.

Geography teachers can use the structured thinking offered by the capabilities approach when curriculum making, and it is for this reason that Lambert (2016) describes GeoCapabilities as a conceptual 'framework' for teachers. The framework ensures powerful subject knowledge is at the heart of good geography curriculum and that powerful pedagogies (Roberts 2014) are the means by which teachers can enable children to engage with geographical knowledge to develop GeoCapabilities. It enables teachers to ensure the moral dimension of the subject is grounded in its knowledge context so as to avoid 'morally

4 The 'Capabilities Approach' to Geography Education

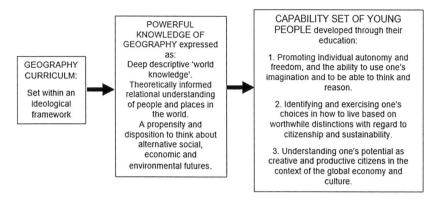

Fig. 4.3 GeoCapabilities as expressed by Solem et al. (2013), in which powerful knowledge (as expressed by Lambert and Morgan 2010) becomes the bridge to connect the curriculum to capabilities development

careless' teaching, an expression of capabilities deprivation. It ensures children have a futures dimension, actively thinking about the choices they will be making about how to live and interact in an ever-changing world. This in turn enables the educated person to have choices about how to live, how to function (including occupational functionings) and to be active agents in the modern world.

Conclusions

The ideas of the capabilities approach to education were developing at the same time as notions of powerful knowledge, and what GeoCapabilities achieves is the combination of both of these ideas. Powerful knowledge articulates what should be taught in a knowledge-led school curriculum but does not express clearly enough the reasons behind this. The capabilities approach links a school curriculum to broad educational aims about living freely and making informed choices in life. This framework was missing the significance of the role that knowledge plays in this process. So GeoCapabilities is a means to express the role geographical knowledge plays in developing the educational capabilities of young people. It is able to marry together the two major concepts of powerful knowledge and the capabilities approach.

> **Questions to Consider:**
> 1. Should societies list specific capabilities that are universally desirable (as Nussbaum encourages) or does this weaken the whole framework that capabilities thinking entails (as Sen believes)?
> 2. What might your list of educational capabilities look like? You can relate this to your thoughts on the aims of education.
> 3. In creating a list of educational capabilities, which of them are developed through skills and which rely on subject knowledge?
> 4. How might the considerations of subject-specific capabilities and skills manifest themselves in a curriculum; what would a school curriculum be like if developing these capabilities and how does it relate to the F1, F2, F3 heuristic?

References

Alkire, S. (2002). *Valuing Freedoms: Sen's Capability Approach and Poverty Reduction.* Oxford: University Press.

Alkire, S., & Black, R. (1997). A Practical Reasoning Theory of Development Ethics: Furthering the Capabilities Approach. *Journal of International Development, 9*(2), 263–279.

Bates, R. (2004). *Developing Capabilities and the Management of Trust: Where Administration Went Wrong.* Paper presented to the Australian Association for Research in Education Annual Conference, Melbourne, December 2004.

Boyes, K., & Watts, G. (2009). *Developing Habits of Mind in Secondary Schools.* Heatherton, VIC: Hawker Brownlow.

Bozalek, V. (2004). *Recognition, Resources and Responsibilities: Using Students' Stories of Family to Renew the South African Social Work Curriculum.* Unpublished Doctoral thesis, Utrecht University.

Clark, D. (2003). Concepts and Perceptions of Human Well-Being: Some Evidence from South Africa. *Oxford Development Studies, 31*(2), 173–196.

Clark, D. (2005). *The Capability Approach: Its Development, Critiques and Recent Advances.* A working paper of The Global Poverty Research Group. Retrieved November 2010, from http://www.gprg.org/pubs/workingpapers/pdfs/gprg-wps-032.pdf

Clark, D., & Qizilbash, M. (2002). Core Poverty and Extreme Vulnerability in South Africa. *Discussion Paper No. 2002–3.* Retrieved July 2011, from http://www.geocities.com/poverty_in_southafrica

Davitt, T. E. (1968). *The Basic Values in Law: A Study of the Ethico-Legal Implications of Psychology and Anthropology*. Philadelphia: American Philosophical Society.

Deprez, L., & Butler, S. (2001). *The Capabilities Approach and Economic Security for Low Income Women in the US: Securing Access to Higher Education Under Welfare Reform*. Paper presented at the First International Conference on the Capabilities Approach: Justice and Poverty, Examining Sen's Capability Approach, St Edmunds College, Cambridge, September 2001.

Finnis, J. (1980). *Natural Law and Natural Rights*. Oxford: University Press.

Flores-Crespo, P. (2001). *Sen's Human Capabilities Approach and Higher Education in Mexico. The Case of the Technological University of Tula*. Paper presented at the First International Conference on the Capabilities Approach: Justice and Poverty, Examining Sen's Capability Approach, St Edmunds College, Cambridge, September 2001.

Galtung, J. (1980). *The True Worlds: A Transnational Perspective*. New York: Free Press.

Gasper, D. (2004). *The Ethics of Development*. Edinburgh: University Press.

Giddens, A. (1984). *The Constitution of Society: Outline of the Theory of Structuration*. Oxford: Polity Press.

Griffin, J. (1986). *Well-Being: Its Meaning, Measurement and Moral Importance*. Clarendon Press.

Hinchliffe, G. (2006). *Beyond Key Skills: Exploring Capabilities*. Presentation given on 16 June 2006 to Networking Day for Humanities Careers Advisers in London. Retrieved September 2010, from http://www.google.co.uk/url?sa=t&rct=j&q=&esrc=s&source=web&cd=1&cts=1330791142639&ved=0CCYQFjAA&url=http%3A%2F%2Fwww.english.heacademy.ac.uk%2Fadmin%2Fevents%2FfileUploads%2FBeyond%2520Key%2520Skills%2520-%2520Exploring%2520Capabilities%2520(HUM).ppt&ei=2UJST5mPMdTY8QPOb3ybBQ&usg=AFQjCNFb2r8hw4htPWDWiZEADCclXF3yJA

Hinchliffe, G. (2009). Capability and Deliberation. *Studies in Philosophy and Education, 28*, 403–413.

Hinchliffe, G., & Terzi, L. (2009). Introduction to the Special Issue 'Capabilities and Education'. *Studies in Philosophy and Education, 28*, 387–390.

Hulme, M. (2008). Geographical Work at the Boundaries of Climate Change. *Transactions of the Institute of British Geographers, 33*, 5–11.

Iceland, J. (2004). *Income Poverty and Capability Deprivation: How Strong Is the Association?* US Census Bureau. Retrieved December 2010, from http://cfs.unipv.it/ca2004/papers/iceland.pdf

Jackson, P. (2006). Thinking Geographically. *Geography, 91*(3), 199–204.
Lambert, D. (2011a). Reviewing the Case for Geography, and the Knowledge Turn in the National Curriculum. *The Curriculum Journal, 22*(2), 243–264.
Lambert, D. (2011b). Reframing School Geography: A Capability Approach. In Butt (Ed.), *Geography, Education and the Future*. London: Continuum.
Lambert, D. (2016). Geography. In D. Wyse, L. Hayward, & J. Pandya (Eds.), *The Sage Handbook of Curriculum, Pedagogy and Assessment*. London: Sage Publications.
Lambert, D. (2017). *GeoCapabilities*. Presentation given at the Geography Teacher Educators Conference, University of Plymouth, 27–29 January 2017.
Lambert, D., & Morgan, J. (2010). *Teaching Geography 11–18 a Conceptual Approach*. Maidenhead: OUP.
Lambert, D., Solem, M., & Tani, S. (2015). Achieving Human Potential Through Geography Education: A Capabilities Approach to Curriculum Making in Schools. *Annals of the Association of American Geographers, 105*(4), 723–735.
Lamsal, M. (2012). The Structuration Approach of Anthony Giddens. *Himalayan Journal of Sociology and Anthropology, 5*, 111–122.
Lasswell, H. D., & Holmberg, A. R. (1969). Toward a General Theory of Directed Value Accumulation and Institutional Development. In R. J. Braibanti (Ed.), *Political and Administrative Development*. Durham, NC: Published for the Duke University Commonwealth-Studies Center by Duke University Press.
Maude, A. (2016). What Might Powerful Geographical Knowledge Look Like? *Geography, 101*(1), 70–76.
Nussbaum, M. (1988). Nature, Function and Capability: Aristotle on Political Distribution. *Oxford Studies in Ancient Philosophy*, Supplementary Volume, 145–184.
Nussbaum, M. (1990). Aristotelian Social Democracy. In B. Douglas, G. Mara, & H. Richardson (Eds.), *Liberalism and the Good*. New York: Routledge.
Nussbaum, M. (1995). Human Capabilities, Female Human Beings. In M. Nussbaum & J. Glover (Eds.), *Women, Culture and Development: A Study of Human Capabilities*. Oxford: Clarendon Press.
Nussbaum, M. (2000). *Women and Human Development: The Capabilities Approach*. Cambridge: Cambridge University Press.
Nussbaum, M. (2003). Capabilities as Fundamental Entitlements: Sen and Social Justice. *Feminist Economics, 9*(2–3), 33–59.
Nussbaum, M. (2006). Education and Democratic Citizenship: Capabilities and Quality Education. *Journal of Human Development, 7*(3), 385–395.

Nussbaum, M. (2016). *Not for Profit: Why Democracy Needs the Humanities* (Updated ed.). In *The Public Square* (Updated ed.). Princeton, NJ: Princeton University Press.
Page, E. (2004). *Teacher and Pupil Development and Capabilities*. Work in progress, presented at the Capability and Education network seminar, St Edmunds College, Cambridge, 1 June 2004.
Qizilbash, M. (1996). Ethical Development. *World Development, 24*(7), 1209–1221.
Qualifications and Curriculum Authority (QCA). (2008). *The Big Picture of the Curriculum*. Retrieved November 2008, from http://curriculum.qcda.gov.uk/key-stages-3-and-4/organising-your-curriculum/principles_of_curriculum_design/index.aspx?return=/News-and-updates-listing/News/Teaching-of-new-secondary-curriculum-begins.aspx
Raynor, J. (2004). *Girls' Development and Capabilities in Bangladesh*. Work in progress, presented at the Capability and Education Network Seminar, St Edmunds College, Cambridge, 1 June 2004.
Roberts, M. (2014). Powerful Knowledge and Geographical Education. *The Curriculum Journal, 25*(2), 187–209.
Robeyns, I. (2003). *The Capability Approach: An Interdisciplinary Introduction*. Retrieved March 2011, from http://www.capabilityapproach.com/pubs/323CAtraining20031209.pdf
Robeyns, I. (2005). The Capability Approach: A Theoretical Survey. *Journal of Human Development, 6*(1), 93–114.
Robeyns, I. (2017). *Wellbeing, Freedom and Social Justice: The Capability Approach Re-examined*. Cambridge, UK: Open book Publishers.
Robinson, M. (2013). *Trivium 21c: Preparing Young People for the Future with Lessons from the Past*. Bancyfelin, UK: Independent Thinking Press.
Royal Geographical Society with Institute of British Geographers (RGS IBG). (2016). Geographers Remain among the Most Employable. Retrieved September 2018, from http://news.rgs.org/post/153564709243/geographers-remain-among-the-most-employable
Saito, M. (2003). Amartya Sen's Capability Approach to Education: A Critical Exploration. *Journal of Philosophy of Education, 37*, 17–34.
Sen, A. (1985a). Well-being, Agency and Freedom. *The Journal of Philosophy, 82*(4), 169–221.
Sen, A. (1985b). *Commodities and Capabilities*. Amsterdam: North Holland.
Sen, A. (1987). The Standard of Living. In G. Hawthorn (Ed.), *The Standard of Living*. Cambridge: Cambridge University Press.

Sen, A. (1999). *Development as Freedom.* Oxford: University Press.
Sen, A. (2004). Capabilities, Lists and Public Reason: Continuing the Conversation. *Feminist Economics, 10*(3), 77–80.
Sharp, J., & Watts, M. (2004). *Go Tell It on The Mountain: The Value of RE Beyond School.* Paper presented at the British Educational Research Association Conference, University of Manchester, 2004.
Solem, M., Lambert, D., & Tani, S. (2013) Geocapabilities: Toward an International Framework for Researching the Purposes and Values of Geography Education. *Review of International Geographical Education, 3*(3). Retrieved December 2013, from http://www.rigeo.org/vol3no3/RIGEO-V3-N3-1.pdf
Stewart, F. (2001). Women and Human Development: The Capabilities Approach by Martha C. Nussbaum. *Journal of International Development, 13*(8), 1191–1192.
Terzi, L. (2003). *A Capability Perspective on Impairment, Disability and Special Needs: Towards Social justice in Education.* Paper presented at the Third International Conference on the Capability Approach, University of Pavia, September 2003.
Terzi, L. (2004). *On Education as a Basic Capability.* Paper presented at the Fourth International Conference on the Capability Approach. University of Pavia, September 2004.
Terzi, L. (2005). *Equality, Capability and Justice in Education: Towards a Principled Framework for a Just Distribution of Educational Resources to Disabled Learners.* Paper prepared for the 5th International Conference on the Capability Approach: Knowledge and Public Action. Paris 11–14 September 2005. Retrieved February 2011, from https://www.researchgate.net/profile/Lorella_Terzi/publication/253934371_Equality_Capability_and_Justice_in_Education_Towards_a_Principled_Framework_for_a_Just_Distribution_of_Educational_Resources_to_Disabled_Learners/links/0deec5304cc288b1cd000000.pdf
Walker, M. (2006). *Higher Education Pedagogies. The Society for Research into Higher Education.* Maidenhead: Open University Press and McGraw-Hill.
Watts, M., & Bridges, D. (2003). *Accessing Higher Education: What Injustice Does the Current Widening Participation Agenda Seek to Reform?* Paper presented to the Transforming Unjust Structures, Capabilities and Justice, Von Hugel Institute, St Edmunds College, Cambridge, June 2003.
White, J. (1973). *Towards a Compulsory Curriculum.* London: Routledge & Keegan Paul.
Young, M., & Muller, J. (2010). Three Educational Scenarios for the Future: Lessons from the Sociology of Knowledge. *European Journal of Education, 45*(1), 11–27.

5

Developing GeoCapabilities: The Role of Research

Introduction

Teaching seems increasingly influenced by educational research, with many roles in schools now dedicated to the undertaking and dissemination of research findings. The ideas discussed in this book, GeoCapabilities, the Future 3 (F3) curriculum and ideas of powerful knowledge, are all conceptual, a means to consider the curriculum and the place of knowledge within it. Yet the ideas did not develop purely in isolation. Research has played an important role in developing and honing the concepts, and this chapter explains the ways various research has added not only rigour to the ideas but a confidence in the approach to curriculum thinking it offers.

This book so far has set up a series of propositions and arguments about the nature of curriculum, and the role of subject knowledge within it. Although the arguments are based around fully referenced ideas, it could be easy to dismiss the actual notion of GeoCapabilities as being purely conceptual, with no grounding in the reality of everyday schooling. This was certainly the case in 2010 when ideas around the capabilities approach to education, powerful knowledge and GeoCapabilities

© The Author(s) 2019
R. Bustin, *Geography Education's Potential and the Capability Approach*,
https://doi.org/10.1007/978-3-030-25642-5_5

were in their infancy. What followed was my own doctoral research and two fully funded international research projects, which helped inform and hone the concept. A third project is now underway to build and develop the ideas from the first two. This chapter tells the story of this research, its outcomes and the ways it has influenced the discourse on GeoCapabilities.

Developing GeoCapabilities Research: 'Engaging Places' Curriculum Development Project and Bridging the 'Great Divide'

Research into the curriculum and the place of knowledge within it is certainly not new and much of it helped lay the foundations for what would become powerful knowledge and capabilities debates of the 2010s. My own thinking on these issues began as a young teacher undertaking master's level research. I chose to find a means to bridge the 'great divide' between school and university level geography that existed at the time, something that was a particular interest to me as explained in Chap. 3. I had been fascinated by developments in the academic study of urban geography, especially the sociological turn which engaged in the lived spaces of people in cities, based on Ed Soja's (1996, 2000) conceptions of First, Second and 'Thirdspace'. In Soja's (2000) work the Firstspace is the built environment, the Secondspace is the imagined, representational space and this combines in the Thirdspace to create a fully lived experience for people in places. I taught this through three cycles of action research with pupils in schools to counter what I perceived to be the uninspiring teaching of urban geography through outdated land use models, which themselves were based on Industrial cities, not complex post-industrial metropolises. This research was published in a number of sources (e.g. Bustin 2011a, b) and with the A Level reforms of 2016 which encouraged a bridging of university and school geography, the work was revisited and updated (e.g. Bustin 2019). This research formed part of the materials produced for the 'Engaging Places' curriculum development project, led by the Geographical Association, which was funded by the British Government's Department for Culture Media and Sport and CABE, the Commission for Architecture and the Built Environment. This project aimed to provide classroom materials across a number of subjects to help children develop an interest in architecture, buildings and cities.

Despite the positive interest in the work, and the exciting developments across a number of subjects from 'Engaging Places', when the research was finished there were still many questions about the place of geographical knowledge in education and the nature of the school curriculum. It was around this time that Professor David Lambert was developing his thoughts about GeoCapabilities and the role of powerful knowledge. He was my tutor for my master's research and in conversation with him at the start of the 2010s, he made me realise that an answer to some of my frustrations may lie in the ideas he was developing. At this time, he was developing a project to explore his ideas and this became known as the 'GeoCapabilities 1' project.

GeoCapabilities 1 Project

The first GeoCapabilities project ran from 2012 to 2013 and was led by the Association of American Geographers and funded by the US National Science Foundation's Geography and Spatial Science program. The project itself is explained in detail in Solem et al. (2013). This was the first attempt to apply the ideas of the capabilities approach to geography education. The project itself involved two stages of original research using data from three countries: USA, England and Finland. Outlines of their countries' educational systems are given in Chap. 1 of this book; indeed their inclusion in this book is due to the focus they received during this project. I was not involved personally in this project and it was relatively small scale.

The project asked two research questions:

1. In what ways do national geography standards in the US, England and Finland portray the subject as a "powerful knowledge" in relation to human capability development?

2. In what ways is the capabilities approach potentially helpful in shaping approaches to curriculum making and developing teachers as leaders in schools? (Solem et al. 2013, p. 221)

To answer these, the project team chose three of the capabilities that Martha Nussbaum (2000) articulated when she expanded on Sen's (1980) framework. These were selected due to their natural applicability to geography. These were presented in Chap. 4 in Table 4.4. The wording of the capabilities was changed slightly to make a more explicit link to geography.

The article, Solem et al. (2013), also set clear parameters for the research by stating what the research was not setting out to achieve. As they argued:

- We are not defining a universal rationale and justification for geography education.
- We are not proposing international standards for geography education.
- We are not advocating a universal approach to teacher preparation in geography. (ibid., p. 220)

This project was intentionally small scale and narrowly focused. The project team looked at how the printed curriculum documents provide opportunities for young people to develop these three specific capabilities. The data was derived solely from the printed materials and involved coding of the curricular documents. A link was then made to how this has implications for curriculum making with some real-life examples of classroom activities that could enable students to enhance these capabilities. This was tested in front of a workshop of teachers, educators and academics at the Eurogeo conference in Bruges, Belgium in 2013. Table 5.1 is taken directly from Solem et al. (2013) and identifies the key outcomes from the project.

The GeoCapabilities 1 project had many benefits for the understanding of the concept of GeoCapabilities. It was the first attempt to specifically link geography as a school subject to a broader set of capabilities as expressed in the literature. As Table 5.1 shows, this was done with relative success. Geography as a school subject does have a value beyond simply learning content for its own sake, and capabilities seems to be a means to

Table 5.1 Outcomes of the GeoCapabilities 1 project from Solem et al. (2013)

GeoCapabilities	Synthesis findings (USA, Finland, England)	Implications for curriculum making (examples)
Promoting individual autonomy and freedom and the ability to use one's imagination and to be able to think and reason.	A shared view in the standards is that geography education equips individuals with the ability to think and reason using diverse forms of locational data and knowledge of human and natural systems in different (and sometimes unique) place contexts. This contributes to the empowerment of individuals to think critically and creatively, whether independently or in collective decision-making and problem-solving contexts, about change and alternative futures.	Teachers in the USA, Finland and England participate in online projects and discussions to offer diverse examples of how their fellow citizens face decisions on where to live, what to build where, how and where to travel, how to conserve energy, how to wisely manage scarce resources and how to cooperate or compete with others. On the basis of these exchanges, teachers work together to develop curriculum materials that engage students in geographic questions of this nature and demonstrate the significance of context and perspective.
Identifying and exercising one's choices in how to live based on worthwhile distinctions with regard to citizenship and sustainability.	Reform of geography in all three countries is driven by greater attention to the idea of sustainability and mandates for environmental stewardship. Knowledge of human–environment relations is essential for understanding environmental and development issues at local, regional, national and international scales and how individual and collective decisions about the future can be enhanced on the basis of this knowledge.	Teachers in the USA, Finland, and England participate in online exchanges of data on energy consumption based on household energy logs. They interpret similarities and differences in localised decision making using comparable data for developing regions, considering the relevance of urban versus rural land use and energy choices and so on. This experience prepares them to create similar classroom activities for their students and also to engage other teachers in thinking about environmental questions from a comparative perspective.

(continued)

Table 5.1 (continued)

GeoCapabilities	Synthesis findings (USA, Finland, England)	Implications for curriculum making (examples)
Understanding one's potential as a creative and productive citizen in the context of the global economy and culture.	Citizens require geographic knowledge and perspectives on economic processes and conditions in different regions to compete and cooperate effectively in a global market while being mindful of the impact of choices, the diversity of cultural approaches to business and economic decision making, questions of how to act ethically, and the value of considering the greater good.	Teachers in the USA, Finland and England collect sales data on products manufactured under a variety of trade relationships between their nations and developing regions, considering and debating the costs and benefits to producers and consumers. They then co-develop a list of questions and have their students engage in online discussions about the relative merits of trading systems and how this knowledge might affect their future choices as consumers and business owners.

articulate this. Capabilities becomes the bridge between geographical knowledge and some broader educational aims (Figure 4.3 in Chap. 4 shows how this this idea can be modelled). The specific examples outlined in the second column of Table 5.1 provide teachers with actual examples of how this thinking might be of use in the classroom.

This first project did have a number of limitations which the team themselves were quick to recognise. Firstly, it only looked at the curriculum as stated in national standards, which may or may not relate to the actual experiences of geography in classrooms in schools. Further research went on to show there is a considerable difference between what is published nationally and by schools, and what actually goes on in those schools. A critical part of the research that was missing was the impact of the concept on practitioners in schools; this research was at a theoretical level rather than a practical level. Thus, by the end of the one year project, there was a need to expand the empirical basis and this is what inspired the GeoCapabilities 2 project.

Geocapabilities 2 Project

This project ran from 2013 to 2017 and was funded by the European Union 'Comenius' fund. The project, its aims and scope are discussed in detail in Lambert et al. (2015) as well as in the GeoCapabilities website. The project involved nine partners from a range of organisations: The University College London Institute of Education, UK, and the University of Helsinki, Finland, were the 'academic' university departments involved; the American Association of Geographers were an external partner (not being from the EU), who along with the Geographical Association and Eurogeo were the subject association partners; a number of schools in Finland and Greece, and two in the UK, were also involved. In addition, a large number of 'associate' partners joined the project—individuals and organisations throughout the world who were keen to share and take part in the work.

The GeoCapabilities 2 Comenius Project aimed to:

- Examine the potential of the capabilities approach to improving the professional development of in service teachers and student teachers.

- Develop a GeoCapabilities teacher training course, with international reach and credibility to make capabilities more visible under the LLLP (lifelong learning platform).
- Develop a Web portal for the establishment of technology-mediated communication and collaboration between teachers and trainees
- Produce a curriculum making methodology that provides enhanced leadership potential for teachers in their own classrooms, and for middle management teams as curriculum developers in schools and other settings. (From the GeoCapabilities project interim report, p. 7, accessed 2017)

The work of the project involved three distinct stages. Initial meetings had an academic focus, to ensure all partners had a clear understanding of the concepts underpinning GeoCapabilities including powerful knowledge, F1 (Future 1), F2 (Future 2) and F3 curriculum, capabilities, and curriculum making. This was the 'conceptualisation and research' phase. This took time, particularly in a multilingual context where many of the ideas were difficult to translate, or where specific words had different meanings across Europe such as 'didactics'; in Europe this term has similar aspirations to the notion of curriculum making, but in the English context didactic teaching refers to a direct form of pedagogical instruction which involves little input from the pupil. The second stage, 'critique, evaluation and design', involved the project team designing the teacher training materials, the web-based platform which would host it, and downloadable content that would form the main output of the project. As part of this, a small number of workshops were undertaken at various educational conferences throughout the world. The final section was 'implementation, dissemination and exploitation', which involved lectures, workshops, teacher training events throughout the world in which partners shared the outcomes of the project, and the ideas of GeoCapabilities. School partners played a major role here, hosting local training events for their own staff and those in local schools. Further training, workshops and lectures took place throughout Europe and the USA as well as Australia, China, and Singapore.

The main project output included a website, (www.geocapabilties.org), and a social media presence (e.g. @Geocapabilities Twitter account),

5 Developing GeoCapabilities: The Role of Research   139

which contains information, video clips, links to papers and chapters about GeoCapabilities and its founding concepts. It also contains a range of teacher training materials aimed at teacher educators. Four modules of training can be done online or downloaded to be used in a variety of settings. These are:

Module 1: The capability approach and powerful disciplinary knowledge
Module 2: Curriculum making by teachers
Module 3: Video case studies
Module 4: Curriculum leadership and advocacy

The project and subsequent materials attempted to empower teachers to be 'curriculum leaders' and to take responsibility for the nature and knowledge content of their own geography curricula and this is the ethos of the project and this book.

Reviewing the project, it was successful as it was able to reach a large number of educational professionals who engaged and questioned the materials. The international audience of the research showed there is a universality about many of the ideas. The website has been used, and the training materials engaged with and the Twitter feed gained followers. The major challenge was in the translation of the key concepts and although there is a glossary on the website, the extent to which all those engaging with the website have a full understanding of these concepts is questionable. This becomes particularly problematic as although the training materials can be engaged with directly by interested individuals it is envisaged that they will be given to students on a course by a course leader. How that course leader uses the materials is beyond the control of the project team.

I was actively involved in the GeoCapabilities 2 project as the lead staff member from partner 8, a UK secondary school where I worked as Head of Geography. This meant I was actively involved at the heart of project discussions, which ran concurrently with my own doctoral research into GeoCapabilities. The project was therefore able to influence my thinking, and I was able to contribute my own ideas back to the project, particularly as many of the early project meetings were spent unpicking key concepts and I was able to offer my understanding of these having already done a review of the existing literature. As a school partner, my practical involvement in the project was more in the latter stages, where the school hosted teacher training opportunities. Another key similarity between my own research and the projects is the role of Professor David Lambert at UCL Institute of Education, who was both supervising my research and leading the research projects.

GeoCapabilities 3 and 'Powerful Geography' Projects

The completion of the GeoCapabilities 2 project in 2017 spawned a number of initiatives which use and develop the ideas further. A third GeoCapabilities project began in October 2018, led by Dr David Mitchell at the UCL Institute of Education. This new project has a focus on social justice in education, finding a practical means to realise the potential of a geography education. As the pages of the GeoCapabilities website outline:

GeoCapabilities 3 seeks to answer 2 main questions:

1. Is there a social justice dimension to GeoCapabilities? and:
2. How can a GeoCapabilities approach benefit schools (teachers/pupils) in challenging (socio-economic) circumstances towards the goal of 'powerful knowledge for all? (Geocapabilities.org, accessed January 2019)

The project aims to work closely with teachers and schools in challenging socio-economic areas across Europe to use the principles of powerful knowledge, within the framework of the capabilities approach to develop high-quality curriculum materials to use in schools in these areas.

In the USA, the National Centre for Research in Geography Education has launched a major project entitled 'Powerful Geography', taking its title from 'powerful knowledge'. The work of the project involves research-led initiatives to inform teacher training in various US States to support the development of the school geography curriculum. The project also held a 'Powerful Geography' conference in November 2018 in San Jose, Costa Rica, in association with the University of San Jose's geography department. Materials from the conference, including a video of David Lambert's keynote lecture, are available online (e.g. geocapabilities.org, powerfulgeography.org).

Doctoral Research: An Investigation into GeoCapability and Future 3 Curriculum Thinking in Geography

Running alongside the second of the GeoCapabilities projects was my own part-time doctoral research into GeoCapabilities (2010–2016). The full research elements are explained more thoroughly in the final thesis (Bustin 2017); the purpose of this section is to give a brief overview of how the research was conducted and, more significantly, how it helped develop the concept of GeoCapabilities.

The research was deliberately different to the work of the projects. The focus was one school in detail and to look at the extent to which notions of the capabilities approach and Future 3 thinking could be of use to teachers. The overall research question was:

How useful is geocapability as a framework for Future 3 curriculum thinking in geography?

This main enquiry question was subdivided into three further research questions:

1. *How do the 'structural features' of education promote curriculum making in geography?*
2. *How can capability develop student agency?*
3. *What contribution does geographical knowledge make to the development of capability?*

The data collection involved a series of in-depth interviews with geography teachers, school leaders, parents, governors and group interviews with pupils. These were conducted in one main case study school and a further school which acted as a means to triangulate some of the findings. The interviews asked a range of questions to understand more about the nature of education and the role of teachers. These were 'semi-structured', which enabled a more conversational style to develop in which attitudes and opinions could be elicited clearly. Table 5.2 gives a couple of examples of what the interview process was revealing.

Table 5.2 Topics of conversation in the semi-structured interview process for two groups of interviewees (from Bustin 2017)

Interviewee	Topic of interview
School leadership team and governors	Their views about school geography, based on personal recollection - Their views on the purpose of education—the relationship between subjects and broader educational goals, and the extent to which the former affects the latter, and how this is evident in the school. - Their understanding of the school ethos, and how it manifests itself. - How geography as an academic subject fits into the wider school curriculum. - How decisions are taken about which subjects to include and exclude from the school curriculum and the role of extracurricular provision in the school.
Geography subject specialists	Their views about school geography, based on personal recollection. - Their views on the purpose of education, and the role that subjects plays within this, and geography as a discrete subject. - A reflection on the nature of the geographical knowledge they learnt at school and that which they now teach, reflecting on similarities and differences. - How they make the geography curriculum; how they mediate between the differing pressures of the needs of the pupils, national curriculum, exam boards, textbook series, facilities and learning resources available and their values of geography education. - Their thoughts about the school facilities and how it affects pupil learning.

Including 2 pilot interviews, 16 in-depth interviews were conducted, each lasting about an hour, and 11 group interviews involving over 50 student voices. They were recorded on a voice recorder and then typed to produce a transcript.

The interview data was analysed through content analysis, using the coding technique from Gibson (2013), who recommends identifying a series of a priori then post priori codes in the text. These are labels that can be used to identify particular words or themes in the text. The initial set of a priori codes were based on initial understanding of the data. The text was highlighted according to a range of codes using specialist research

5 Developing Geocapabilities: The Role of Research

software. One such code was for example, 'National Curriculum', which was used each time an interviewee discussed the National Curriculum. Having completed this exercise, many of the codes chosen were too broad, and others too narrow and a new set of what Gibson (2013) describes as post priori codes were identified and the transcripts re-coded. At the end of the process, there were 29 post priori codes identified in the text, and these were then classified into 5 themes; aims of education, structural features of curriculum, subject disciplines, control of education and school geographical knowledge.

The second stage of data collection involved a more practical 'curriculum making workshop' with the geography teachers in the case study school in which they were asked to plan a sequence of lessons. The purpose was to see how their ideas about teaching and education manifested themselves in a practical activity which focused on what was being taught in a sequence of geography lessons. The workshop involved the team of geography teachers planning an overview for a sequence of approximately six weeks of lessons on 'Russia'. This topic was chosen as it was part of the 2013 National Curriculum and thus had to be covered in schools; but at the time of the research, it was not yet part of the case study school's curriculum. Therefore, the teachers had to think creatively from the start without being able to rely on already planned materials.

After initial discussions and some ideas about what they might want to include they created a basic overview of their lesson sequence, shown in Fig. 5.1. Their discussions gave an insight into how this was achieved.

The teachers were then given a 'GeoCapability Framework', a table devised for the research to help them check the level of geographical content. At the top of each part of the table was one of the three ways the powerful knowledge of geography has been expressed according to Lambert and Morgan (2010). The teachers had to map their initial ideas into the table. This then allowed them to see any gaps or omissions in the geographical content they had planned. They could add more material into the lesson sequence at this stage in light of this exercise. This was designed to enable the teachers to reflect on the powerful geographical knowledge contained in the lessons they had planned to avoid the trap identified by Roberts (2010) of 'where's the geography?' The final stage of the workshop was a critical reflection of the exercise and what it

Fig. 5.1 The results of the initial discussions about what to include in a lesson sequence on 'Russia' (from Bustin et al. 2017)

identified to those taking part. This practical part of the research was published (Bustin et al. 2017) and the completed GeoCapabilities Framework from the original workshop is shown below, in Fig. 5.2.

To generate the data from the workshop in addition to the completed lesson plans, notes and completed framework, a 'narrative' was written, the story of the workshop in which I reported and critically reflected on what I heard and saw. To analyse this, the 29 post priori codes from the interview data were then applied to the narrative text.

The themes and ideas discussed in interviews and from the workshop narrative enabled a series of 'contentions' to develop from the research. Each of the contentions here is substantiated by a series of quotations taken directly from the interview or narrative data.

Table 5.3 shows these, arranged into five overarching themes.

5 Developing GeoCapabilities: The Role of Research   145

The capability approach: GeoCapability framework		
Deep descriptive 'world knowledge'	Theoretically informed relational understanding of people and places in the world	Propensity and disposition to think about alternative social, economic and environmental futures
Place and space	Human and physical processes	Choices about how to live
Russia — where/extent, cities, physical landscape	Why cities are where?	
Natural resources	Why oil/gas/coal?	Extraction Supply/demand Sustainability Conflict of use
Contemporary conflicts — Ukraine — location EU/Russia divide	Why the conflict — cultural identity? Age Resources?	Future solutions to conflict
World stage — Development Russia/BRICs	Classification/ globalisation rate/human rights/HDI/trade	Sustainability

Fig. 5.2 The completed GeoCapabilities Framework, created for the research (from Bustin et al. 2017)

These contentions, drawn directly from empirical data, became the basis for the further development of the concept of GeoCapabilities. The notions of powerful knowledge and capabilities were used to provide a response to the contentions; in many cases 'capabilities' are able to help resolve or respond to some of the contentions identified. For example, a key contention revealed in the analysis of the interviews was the idea that teachers felt heavily controlled in their curriculum making by external forces such as the National Curriculum and awarding authorities

Table 5.3 The contentions derived from doctoral research into GeoCapabilities

Aims of education:
1. There is much discrepancy between what people regard as being the aims of education.
2. Skills, particularly related to careers, are given a higher status than knowledge when related to the aims of education.
3. Developing a sense of moral responsibility was seen as being an important aim of schooling, but this was a separate consideration to knowledge and skills acquisition. Using separate timetabled citizenship lessons was not considered the best means to develop these ideals.
4. The passing of exams seems to be an implicit and fundamental aim of schooling.
5. The data also revealed that the role and status of knowledge are low; it is seen as a memory test, used to help develop skills.

Structural features of curriculum:
6. The Habits of Mind, a curriculum structure introduced at the main case study school (as discussed in Chap. 2), is widely ignored and disregarded by staff and pupils in the school. Some teachers have interpreted the demands of the Habits of Mind too literally.
7. The beliefs and perceived nature of the children and teachers are seen as an important starting point for curriculum making by teachers.

Power and control of education:
8. There is a series of structural features all perceived to be imposing power over the teacher, and these are as much about the physical environment as about key stakeholders, such as the government and examination awarding authorities.

Subject disciplines and the curriculum:
9. Subjects are still regarded as the basis of a curriculum organisation in schools, but are defined in terms of a skill set, not by knowledge content.
10. The articulation for the maintenance of a subject-based curriculum was less well expressed.
11. Examination grades only have a limited lifespan of usefulness.

School geographical knowledge:
12. What is regarded as 'geography' is ill defined and problematic. The three-part expression of geography's powerful knowledge (from Lambert and Morgan 2010) was challenged.
13. There seems to be a variety of ways of structuring geographical knowledge for teachers to assist in curriculum making; teachers in the workshop had a belief in the importance of contemporary issues driving the geography curriculum.
14. The use of the GeoCapabilities Framework created for this research can assist teachers in ensuring a balance between place, process and future's geographical knowledge.

(contention 8). This was evidenced by a number of voices in the interviews. In the capabilities discourse this control is part of the 'structural features' of curriculum (see the discussion in the previous chapter on this). Yet the teachers in the workshop designing the lessons on 'Russia' behaved differently to a group of teachers feeling heavily controlled, instead getting visibly excited about the lessons they were creating. For those teachers, contemporary issues was the most significant factor that drove their curriculum making (contention 13) and this structured the knowledge of the lessons and the lesson sequence they created (as shown in Fig. 5.1). The notion of contemporary issues based geography was not an external factor driving curriculum but something from the teachers themselves, from their own experience and their own ideas about what is important in the geography curriculum. The teachers had ignored the 'heavy' influence of National Curriculum requirements, awarding authorities and other structures and planned the lessons free of those controls. The Habits of Mind, despite offering a big picture of curriculum in the case study school, and something the teachers identified in interview as a controlling influence, was completely ignored in the workshop (this forms contention 6). This analysis led to some important considerations about powerful knowledge and curriculum control. Teachers were able to exercise 'choice' over knowledge when curriculum making. Some of the overarching structural features of curriculum gave the teachers a set of parameters to work within; they still needed to plan lessons that were a set length of time and a set number; they did not have choice over the topic of 'Russia' as this was set for them; they did not have control over the year group the lessons were aimed at. Yet within these parameters, they had free control over what they chose to teach and were able to exercise their own ideas and ideologies. It was this analysis that led to the discussions around 'structuration theory' in the previous chapter; teachers can use or subvert the influence of the structural features of curriculum to create the lessons they want to for their students. This analysis in turn highlighted the significance of teacher choice in the process of curriculum making and in part why the GeoCapabilities 2 project spoke of "teachers as curriculum leaders". The significance placed on teachers to be able to plan around and mediate a range of curriculum pressures leads into the larger debate about the training and preparation of teachers. In order to be able to make the curriculum successfully, based on powerful geographical knowledge, teachers need to be fully trained subject specialists. This particular discussion here provided a response to the first of the three research sub questions about the role of structural features influencing curriculum making decisions.

The second of the research questions asked how capability can develop student agency, part of the 'outcomes' rather than 'outputs' of education. The analysis of the data identified that the outputs were given much more significance than the outcomes in the attitudes of those in this research (contention 4). Achieving examination grades was seen as the aim of education, albeit implicitly. There does need to be a means by which to differentiate between the abilities of young people, particularly in a competitive jobs market. This was deemed an important consideration of the teachers in this research (contention 2). Yet the analysis also identified that examination grades have a limited usefulness in the workplace (contention 11) and this has informed the judgement that whilst exam grades are significant, the emphasis placed on them by those in this research was too great. There is a greater purpose to effective curriculum making, which can be articulated by 'outcomes'. The idea that education is more than exam grades was identified in the analysis of the data with regard to the importance of a moral responsibility (contention 3). This is an expression of a holistic, pupil focussed outcome which can be encapsulated by capabilities. As a result of their education by subject specialist teachers, a young person develops a capability set. This set is able to articulate the ways that person thinks about the world, their understanding, their values and their attitudes towards issues. A capability set cannot be easily measured; it is intangible and is built up throughout education.

The outcomes and outputs alone are not the 'end product' of education. What the capability approach enables is a consideration of the choices that are available to pupils as a result of their education. With a capability set, a young person is able to make choices about how to live, and it is these choices that determine their future. The capability approach identifies two main choices, functioning and agency—terms derived from the literature, explored in the interviews and explained in the last chapter. Functioning attempts to collate the various 'beings' and 'doings' of life, which includes choices about which career or job to take ('occupational functioning', Hinchliffe 2006), itself identified in the analysis as an aim of education (contention 2). The capability approach enables these functionings to be placed within a broader framework of ideas. This notion of a moral education was deemed significant by those in this research (contention 3), with many commenting on the importance of developing a 'moral compass' in pupils. These notions fit well with the idea of agency. 'Agency' provides a means by which the needs of moral education can be expressed as part of the outcomes of education. An example of this can be illustrated through the workshop in the research. The teachers were keen to teach the geopolitical tensions and conflicts that exist between Russia and many of its neighbouring countries. The educated pupils, with this complex knowledge, can then decide how to respond to contemporary Russia, through engaging with national politics and debate.

The third research question referred to the role of knowledge in the capabilities approach. A key contention identified in the analysis was that the status of knowledge was low, akin to a memory test for some pupils and far less significant than skills (contentions 2 and 5). However, the capability approach places a far greater emphasis on knowledge at the heart of a curriculum. Placing knowledge as part of the structural features forces teachers to consider, from the outset, what to teach. The subject specialist geography teachers, driven by a clear justification of why they teach their subject (part of the social structural features), are then able to select from that knowledge in order to inform their curriculum making. The workshop in this research gave an insight into this process. Teachers 'recontextualise' social realist, disciplinary knowledge for pupils in the classroom, using their pedagogic skills to enable pupils to access and develop understanding. This has been articulated in the literature as powerful pedagogies (Roberts 2014), which enable pupils to develop powerful knowledge (Young 2008), and it is this which is an articulation of an F3 curriculum. The analysis of the data, specifically through the role of the teacher workshop, identified the significance of the GeoCapability Framework; focussing on powerful geographical knowledge in the planning stages of a curriculum ensured a rigorous knowledge content to geography lessons (contention 14). For the subject of geography, powerful knowledge has been expressed in three parts; deep descriptive 'world knowledge'; theoretically informed relational understanding of people and places in the world; and a propensity and disposition to think about alternative social, economic and environmental futures. However, the analysis of the data did question the validity of this expression (contention 12).

The importance of knowledge in the capability approach extends beyond curriculum making to consider the reason why young people develop powerful knowledge. The 'powerful knowledge' that pupils engage with through subjects builds up understanding of those subjects and this develops what is being identified as 'knowledge-based capability'. For the subject of geography this is called 'GeoCapability'. GeoCapability includes the powerful knowledge of geography, as expressed through the three-part expression of powerful knowledge, which enables the ability to think like a geographer. The powerful knowledge of geography is not defined as a list of facts; it is ill defined, an assertion from both the literature and the data (contention 12). Schools and external assessment criteria such as examinations test a student's geographical knowledge; much of this is about testing what a pupil can remember, though some questions at A Level do encourage pupils to think geographically by linking people, places and processes. Yet this assessment of outputs suggests a set end point for geographical knowledge, that once a pupil has sat a geography exam their geographical abilities end. GeoCapability, whilst less measurable, is a more long-lasting outcome of a geography education (responding to contention 11) and affects the way a young person sees and interacts with the world. Once formal education ends, the educated person has a knowledge-based capability set, of which GeoCapability forms a part. The significance of knowledge enables pupils to make informed choices about 'functioning' and 'agency'. Powerful knowledge from a variety of school subjects helps young people to make choices and decisions about their place in the world, and react to issues and ideas. This discussion responds to the contention from the analysis that citizenship ideals are best engaged with through traditional subjects (contention 3), where pupils can base life choices on powerful subject knowledge gained through education, rather than through a potentially knowledge-less 'morally careless' curriculum.

This level of analysis of the data informed the ideas discussed in the previous chapters and has led to the 'vision' for a Future 3 curriculum presented in the next chapter.

How Useful Is Geocapability As a Framework for Future 3 Curriculum Thinking in Geography?

The overall research question revolved around the idea of 'usefulness' of GeoCapabilities as a conceptual framework for a Future 3 curriculum. The usefulness of the capability approach to curriculum thinking can be considered in terms of a number of audiences. One broad audience is the vast range of educationalists working in schools. One consideration from the analysis of the data was that there were many different articulations of aims of education (contention 1). The conceptual understanding of capability, based on this contention, is clear in its response to this. With the curriculum thinking offered by the capability approach, the purpose of education is to enable a young person to be free to think, to be, to do and to live life in the way they choose. This is achieved through the development of powerful subject knowledge to develop knowledge-based capabilities. This aim is not derived directly from the needs of examinations, universities, societies or the jobs market, although these have an influence, but from individual young people living the life they choose in the twenty-first century. This clarity could be useful for teachers as well as school leaders.

The capability approach is useful for teachers, as it enables them to think about and to focus on what they are teaching and why they are teaching it. For teachers of geography, the idea of what counts as geographical knowledge can be problematic (contention 12), so the practical Framework of powerful geographical knowledge ensures a geographical knowledge content in the curriculum without dictating its exact contents. The usefulness is in the curriculum planning stages, when courses are being designed that will span more than one lesson; capability is something developed over a sequence of lessons and so teachers can ensure that lessons are developed that will maximise powerful knowledge engagement. For teachers of geography a sequence might involve a mixture of place- and topic-based material over two or three lessons, with students engaging with a variety of data before considering an issue based on their understanding. This issue can then be considered from a variety of alternative viewpoints suggesting alternative futures. This is how the teachers in the workshop designed their sequence on Russia, within the structure offered by the capability Framework. It is the practical Framework that has the potential to provide a means to enable an F3 curriculum. The teachers in the workshop saw the relevance of the curriculum thinking structured by this Framework (contention 14).

The usefulness for teachers is also in its flexibility as a concept. Neither the capability approach nor an F3 curriculum lists what capabilities actually are in practice, nor do they list the powerful knowledge on which that capability is based. This is in a similar vein to the ideas of Sen (1980) who never advocated a list of development capabilities. What the capability approach does advocate is a subject-based curriculum, with subject specialist teachers who are able to work with the subject and pursue 'better knowledge'. By creating a definitive list of what counts as powerful knowledge, it becomes static and something to be learnt rather than an idea to engage with, which can lead to an F1 curriculum. It is the role of the teacher working in specific settings to choose and present the powerful knowledge of their subject to their students. Different teachers working in different schools will have different ideas about what constitutes powerful knowledge, and the capability approach respects these differences as long as the teacher is a subject specialist and is working within their subject discipline. Thus, the capability approach is empowering for teachers. It is trusting of their professional abilities and their understanding of both their subject and education. In this sense, capability forms part of the professional ethos of teachers' work. It is part of the way teachers see themselves and their role as educators. The workshop and its subsequent analysis identified the ways that the geography teachers in this research made these curricular decisions.

The usefulness of the capability approach can also be considered from an educational leadership perspective, both in schools and in wider society. The capability approach as well as an F3 curriculum require subject specialists to be teaching a subject-based curriculum. This was a position

5 Developing GeoCapabilities: The Role of Research

supported by almost all interviewed (contention 9). Yet neither of these two requirements is consistent in all schools, as some of the ideas in Chap. 1 illustrated, and in an increasingly fragmented educational landscape, a set of principles for rigorous curriculum design could be useful. Capability could provide some of those principles. For school leaders, the capability approach demands that teachers need to be teaching the subject in which they have a specialism. If they are not specialists, they will not be best placed to develop knowledge-based capabilities in those children they teach. This is because powerful knowledge requires an immersion in a discipline and that cannot be offered by a non-specialist. This argument also champions the need for teacher training to be based predominantly in universities, with groups of subject specialist trainees working together to learn their profession and reflecting critically on subject knowledge together, rather than training individually in isolation in schools where the time for this in-depth reflection with other specialists would be more limited.

For local and national educational authorities, schools need to pursue a subject-based curriculum, so children will be able to develop knowledge-based capabilities. Without knowledge-based capabilities, young people cannot develop a full capability set to enable choices in life. A lack of access to a subject-based education can be considered a form of 'capability deprivation'.

The specific outcomes of the doctoral research were two-fold. Firstly, the contentions identified were able to develop GeoCapabilities as a concept and revealed a number of significant features, as the example above about teacher choice and control illustrates. These findings were fed into the GeoCapabilities 2 project as it developed and in turn were used in workshops to help further refine the ideas. Whilst still conceptual, the framework of thinking it offers benefitted from this empirical basis. The framework was modelled as part of this research and this forms the 'vision' of a capabilities curriculum presented in the next chapter.

Secondly, the research created a practical tool for teachers to help curriculum making. The practical GeoCapabilities Framework is designed to enable teachers to check the geographical content of their curriculum; using this with teachers subsequent to the research has received a positive reaction.

Future Research

Although the two major funded GeoCapabilities projects—the doctoral research presented here—have now been finished, the idea of GeoCapabilities, powerful knowledge and the Future 3 curriculum are readily and actively being researched. The GeoCapabilities 3 project has begun, and a number of avenues of future research areas are identified here.

The first obvious way to evolve the concept is to broaden the empirical basis of the study. There has only been one in-depth case study analysis of how educationalists working in a whole school might be able to conceive a role for the capabilities approach. Repeating the research by adding voices from a broader range of schools, with different approaches to curriculum organisation and a wider ability range of pupils, would provide evidence to further the concept of capability. How comparable the idea of knowledge-based capability is to all teachers in all schools is something further to study.

Another way to broaden the empirical basis is to change the focus; the focus of all the research into GeoCapabilities so far has been on teachers' work and how teachers conceptualise the curriculum for pupil benefit. Therefore, follow-on research could focus on the pupils themselves, how they respond to and work with notions of powerful knowledge and capability. The use of the practical 'capability Framework' developed in the case study research, to map out aspects of powerful knowledge within topics that the pupils learn, could be a means to investigate pupil perceptions of knowledge. Pupils could be asked to complete a version of this capability Framework for the topics they study in their geography lessons so they can identify the geographical knowledge they have learnt.

A further way to build on this research is through the context of geography education internationally. The terms of reference throughout this book are in an English context (such as the National Curriculum, General Certificate of Secondary Education [GCSE] exams, etc.), although reference has been made to geography education in the USA and Finland to provide a comparison. The GeoCapabilities projects have identified that the capability approach to geography education has a resonance with other geography educationalists in other national settings (e.g. Solem et al. 2013) so one clear further area of study is to develop this idea. The capabilities approach and the notion of powerful geographical knowledge enables geography teachers to see the importance and value of learning geography, and this notion can be translated internationally. The potential importance of GeoCapabilities to provide a rationale and thus an argument for its curriculum inclusion could be significant, but further work is needed.

Finally, the capabilities approach to curriculum thinking can be extended across subject disciplines. The ideas in this book are based on the subject of geography. If the capability approach to education is to be developed and explored as a rigorous curriculum Framework, then further research needs to take place to see if other subject specialists can see a value in its principles. For example the value of powerful knowledge and knowledge-based capability for other subjects could be explored, and it would be of particular interest for subjects which already have a strong curricular 'frame' and more vertical knowledge structures (ideas from Bernstein 2000, as discussed previously) such as maths and physics.

The implications of many of the ideas in this book require school leaders to think holistically about their whole school curricular organisation and, as such, more research needs to look at the role played by leaders to develop a Future 3 vision in schools.

Conclusions

This chapter had a simple aim: to show the ways in which the concept of powerful knowledge, the Future 3 curriculum and GeoCapabilities have developed through research. The ideas presented in this book have an empirical basis to them. This chapter has introduced the main research, but the full details, methodologies and specific research outcomes are available in the various articles published and referenced here. Researching a concept such as GeoCapabilities will always provide empirical challenges. Using teacher voice, or children's voice, is fraught with methodological considerations and the use of one case study, as in the doctoral research, can never give a broad coverage of ideas representing all children.

The ideas developed from the projects, and the doctoral research, are only ever applicable at the time and place in which they were created. If the research was to be repeated, even back at the same school, results and therefore interpretations could well be very different. What this book represents, therefore, is the latest stage in the understanding of a concept that will continue to evolve and develop as more research is conducted. This latest vision for an F3, powerful knowledge-led curriculum is the topic of the final chapter.

> **Questions to Consider:**
> 1. How would you have designed a piece of research to investigate capabilities, powerful knowledge and attitudes towards curriculum?
> 2. How could you critique the research outlined here?
> 3. What are the challenges in investigating a curricular concept like GeoCapabilities?
> 4. Pick one of the future areas of research. How could you design a robust piece of research to investigate it?

References

Bernstein, B. (2000). *Pedagogy, Symbolic Control and Identity: Theory, Research and Critique* (Rev. ed.). London: Taylor and Francis.
Bustin, R. (2011a). The Living City: Thirdspace and the Contemporary Geography Curriculum. *Geography, 96*(2), 61–62.
Bustin, R. (2011b). Thirdspace: Exploring the 'Lived Space' of Cultural 'Others. *Teaching Geography, 36*(2), 55–57.
Bustin, R. (2017). *An Investigation into GeoCapability and Future 3 Curriculum Thinking in Geography.* Unpublished doctoral thesis, UCL Institute of Education.
Bustin, R. (2019). Investigating Lived Space: Ideas for Fieldwork. *Teaching Geography, 44*(1), 17–19.
Bustin, R., Butler, K., & Hawley, D. (2017). GeoCapabilities: Teachers as Curriculum Leaders. *Teaching Geography, 42*(1), 18–22.
Geocapabilities.org Website. Retrieved January 2019, from http://www.geocapabilities.org

Gibson, W. (2013). *Qualitative Data Analysis: Thematic Analysis*. Course Notes from Doctoral Studies Methodology Course, UCL Institute of Education.

Hinchliffe, G. (2006). *Beyond Key Skills: Exploring Capabilities*. Presentation Given on 16 June 2006 to Networking Day for Humanities Careers Advisers in London. Retrieved September 2010, from http://www.google.co.uk/url?sa=t&rct=j&q=&esrc=s&source=web&cd=1&cts=1330791142639&ved=0CCYQFjAA&url=http%3A%2F%2Fwww.english.heacademy.ac.uk%2Fadmin%2Fevents%2FfileUploads%2FBeyond%2520Key%2520Skills%2520-%2520Exploring%2520Capabilities%2520(HUM).ppt&ei=2UJST5mPMdTY8QPO-b3yBQ&usg=AFQjCNFb2r8hw4htPWDWiZEADCclXF3yJA

Lambert, D., & Morgan, J. (2010). *Teaching Geography 11–18: A Conceptual Approach*. Maidenhead: OUP.

Lambert, D., Solem, M., & Tani, S. (2015). Achieving Human Potential through Geography Education: A Capabilities Approach to Curriculum Making in Schools. *Annals of the Association of American Geographers, 105*(4), 723–735.

Nussbaum, M. (2000). *Women and Human Development: The Capabilities Approach*. Cambridge, UK: Cambridge University Press.

Powerful Geography Website. Retrieved January 2019, from powerfulgeography.org

Roberts, M. (2010). Where's the Geography? Reflections on Being an External Examiner. *Teaching Geography, 35*(3), 112–113.

Roberts, M. (2014). Powerful Knowledge and Geographical Education. *The Curriculum Journal, 25*(2), 187–209.

Sen, A. (1980). Equality of What? The Tanner Lecture on Human Values Delivered at Stanford University May 22, 1979. Retrieved November 2015, from http://tannerlectures.utah.edu/_documents/a-to-z/s/sen80.pdf

Soja, E. (1996). *Thirdspace*. Oxford: Blackwell.

Soja, E. (2000). *Postmetropolis*. Oxford: Blackwell.

Solem, M., Lambert, D., & Tani, S. (2013). Geocapabilities: Toward an International Framework for Researching the Purposes and Values of Geography Education. *Review of International Geographical Education, 3* (3). Retrieved December 2013, from http://www.rigeo.org/vol3no3/RIGEO-V3-N3-1.pdf

Young, M. (2008). *Bringing Knowledge Back In: From Social Constructivism to Social Realism in the Sociology of Education*. Abingdon: Routledge.

6

The Potential of a Future 3 'Capabilities' Curriculum

Introduction

This final chapter brings together the body of research that has taken place into GeoCapabilities, powerful knowledge, Future 3 (F3) curriculum thinking and the capabilities approach. In this chapter, the vision for what a Future 3 curriculum could look like in practice is set out, with some practical examples of how curriculum thinking using the capability approach can enable curriculum making in geography. The ideas in this book have been subject to critique and some of the ways the framework has drawn criticism are explained along with a response to the critics. The final section offers some of the challenges that need to be overcome to fully realise the potential of a Future 3 curriculum.

Envisioning a Future 3 'Capabilities' Curriculum

The aspirations of what Young and Muller (2010) identified as the Future 3 curriculum (discussed in Chap. 2) are embedded within the capability approach to curriculum thinking. Young and Lambert (2014) argue that

© The Author(s) 2019
R. Bustin, *Geography Education's Potential and the Capability Approach*,
https://doi.org/10.1007/978-3-030-25642-5_6

F3 "treats subjects as the most reliable tools we have for enabling students to acquire knowledge and make sense of the world" (p. 67). The concept of capability provides a structured way of thinking to enable this. The concept has an educational outcome of pupils using their education to enable choices, choices of functioning and agency. These choices are the product of a well-developed capabilities set made up of knowledge-based capabilities. These are developed through rigorous engagement with the powerful knowledge of subjects, taught by expert subject teachers. These ideals are the key principles of an F3 curriculum—a subject-based curriculum, with powerful knowledge at its heart. Powerful subject knowledge is that created through specialist thought; it is the best knowledge of that subject; it is evidence based but open to debate.

A result of some of the research into GeoCapabilities identified that many teachers and educators were keen to retain a subject-based curriculum but could not articulate why. The capabilities approach to curriculum thinking can provide an articulation. Capabilities links the aims of education, via subjects and powerful subject knowledge, to student outcomes. This gives a framework of curriculum thinking that embodies an F3 curriculum vision. The arguments in this book are not suggesting that the capabilities approach is the only way to achieve an F3 curriculum. What it is arguing is that the capability approach can be one way for teachers to enable an F3 curriculum in schools. Geography, as a school subject taught by geography subject specialists, enables pupils to engage with powerful geographical knowledge, which develops GeoCapability and ultimately leads to geography pupils being able to think in new ways that were not possible before the development of this knowledge. This is an expression of F3 geography curriculum thinking.

In an attempt to bring together many of the various aspects of the capabilities approach to curriculum thinking, Fig. 6.1 shows a conceptual model that explains how the capabilities approach from Amartya Sen and Martha Nussbaum can be applied to thinking about the school curriculum.

Figure 6.1 is a conceptual model that structures teacher thinking within a capability approach. The start of the model is the left-hand side in which there are various 'structural features' that impact the way teachers conceive the curriculum. These are both 'real' and 'perceived' and

6 The Potential of a Future 3 'Capabilities' Curriculum

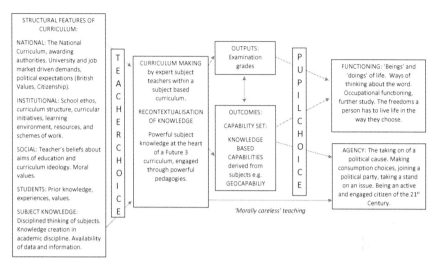

Fig. 6.1 A model of the capability approach to education

provide a structure for the thinking needed for curriculum making. They operate at a variety of scales; nationally they include the National Curriculum, the criteria of awarding authorities and the curricular demands of the national government such as the promotion of various 'public values'. Further structures are provided by the educational institution and take the form of the learning environment, timetable constraints, school ethos and approaches to learning. The teachers themselves also add a layer of structure to curriculum thinking; their own beliefs about education, their values and ideological perspective all affect the way they think and conceive the educational experience. In a similar vein the prior knowledge and experience of the pupils also structures curriculum thinking for teachers who need to ensure their pupils will gain from the lessons, and that their prior experience can be capitalised on to develop new knowledge. Knowledge is modelled as part of the structural features; this includes academic subject knowledge, which for geography would include the way the discipline is conceived in universities, as well as world knowledge available to teachers through the media and their own world experience. Placing knowledge as part of the structural features forces teachers to consider, from the outset, what to teach. The subject specialist geography teachers, driven by a clear justification of why they teach their subject, are then able to select from that knowledge in order to inform their curriculum making.

Teachers are influenced by these structural features when they are involved in curriculum making, which is modelled by the arrows leading to the second of the boxes in the model. These arrows are labelled 'teacher choice' as teachers actively make choices about how to respond to these features; for teachers, some of these features help 'enable' whilst others 'constrain' curriculum decision making similar to the ideals of the 'structuration theory' (Giddens 1984). Whilst the structural features provide an overarching set of parameters within which teachers work, there is still considerable freedom for teachers to think and create freely by actively subverting and working creatively within the enabling structural features. The examples of the teachers' curriculum making in the workshop during the doctoral research described in the last chapter show this process in action.

The second of the main boxes in the model is the process of 'curriculum making'. It is here that teachers enable the curriculum for their pupils through a knowledge-led, subject-based curriculum. It is this part of the model that deals with pedagogy, the actual lesson activities that teachers use to help pupils engage with complex knowledge. These can take many forms; individual lessons would have a set of aims and objectives, learning would be structured to assist engagement with powerful knowledge and the young people would learn. Teachers actively use their knowledge and understanding of their subject to recontextualise it for their students, making it accessible through the ways they frame the materials. It is this skilled activity, the professional work of a teacher, that leads to the development of powerful knowledge and it is for this reason Roberts (2014) talks of 'powerful pedagogies' which drive this process.

The model identifies the purpose of this; effective curriculum making leads to a distinct set of pupil focussed results, shown on the model as 'outputs' of examination grades and 'outcomes' of capabilities. Achieving high examination grades for pupils is a significant aim for teachers and pupils, but research presented in the previous chapter, and some of the anecdotes presented in the first chapter, identifies that this is often seen by teachers as the principle aim of schooling. This output could drive the

6 The Potential of a Future 3 'Capabilities' Curriculum

entirety of the model presented here; it can dominate the curriculum making process and the experience of subjects for pupils. What a capabilities perspective identifies is that whilst examination grades are important for pupils, there is a greater purpose to effective curriculum making, which can be articulated by 'outcomes'. As a result of their education by subject specialist teachers, a young person develops capabilities based on powerful subject knowledge. Students do not learn subjects in isolation in schools. Through engagement with a variety of subjects, students develop a range of knowledge-based capabilities, such as maths capabilities and history capabilities (although more research is needed within these subject education communities to see if capabilities are able to be expressed, and as such if it is a workable framework). This forms an individual's capability set.

This 'set' is able to frame the ways that a young person thinks about the world, their understanding, their values and their attitudes towards issues. A capability set cannot be easily measured; it is intangible and is built up throughout education. The arrow between outcomes and outputs identifies that the two are linked; pupils with a well-developed capability set are more likely to achieve a higher set of outputs, although this is not a given assumption.

Teaching aimed at outputs requires a sole focus on teaching to the test, not straying beyond the examinable materials, spending much class time with practice questions and unpicking various question demands. Pupils become passive recipients of the version of truth that is presented in an authoritative text book as this is what they will be tested on. It runs the risk of creating an F1 (Future 1) curriculum of facts to be learnt and drilled. Teaching aimed at outcomes will not be afraid to stray beyond the syllabus if it helps broaden understanding; it encourages discussion and debate and challenges assertions made in textbooks. Good teachers will balance the needs of both; of course, teachers need to prepare pupils for any examinations but the F3 curriculum vision places a more significant role on pupil outcomes.

Unlike an educational vision which sees the end point of an education as a set of exams, what the capability approach enables, modelled in Fig. 6.1, is a consideration of the choices that are available to pupils as a result of their education. With a well-developed capability set, a young person is able to make choices about how to live, and it is these choices that determine their future. The capability approach identifies two main choices, functioning and agency, terms explained throughout this book. Functioning attempts to collate the various 'beings' and 'doings' of life, which includes choices about which career or job to take. The capability approach enables these functionings to be placed within a broader framework of ideas. Functioning is a holistic term that identifies how a person is able to 'be' in the world. The more developed the capability set, the more choices young people will have available to them, such as the greater the range of career options and university courses that will be accessible to them. If a person has a less developed capability set, their range of choices about how to live is more limited. There is clearly a link to outputs here too; university choices are a product of examination grades as the universities set entrance criteria. But the choice of functioning is directly related to the educational experience of young people, and to the powerful knowledge they have learnt through their subjects. An educated person will have developed a range of knowledge-based capabilities and a variety of skills, and with this they have choices in their lives. They can decide where to live, what work they want to do and where to go on holiday. They can decide to keep on studying or not. Subject knowledge is inherent in the choices over functionings that an educated person has. Through subjects, young people can learn (1) that actions have consequences; (2) about ways of behaving; and (3) about ways to articulate and present coherent ideas and logical arguments, and this will all affect the way a person lives and interacts in the world. The capability approach does not, however, identify strictly subject-specific functionings; it is not true to say that developing GeoCapability leads to purely geographical functionings (if these could actually ever be identified). GeoCapability forms part of a larger capability set on which all functioning can be enabled. Functioning is about the ways a person interacts with society, the freedoms they have in the sense that Nussbaum and Sen envisioned, rather than an insular focus on the ways in which individual minds function.

The concurrent choice available to educated pupils is the idea of the taking on of 'agency'. In capability approach discourse, the notion of agency refers to the abilities of people to act freely, to participate actively in society and to play a full part in political and economic decision making. It does not identify what these decisions would be, but simply that a person with a well-developed capability set would be in a better position to be more active as an agent in society. This is not a radical 'call to arms', but simply making moral and ethical choices about how to live and work in the twenty-first century.

Consideration of agency through the capability approach also identifies the potentially damaging role of 'morally careless' teaching, which teaches young people to take on agency without a consideration of the knowledge underpinning the issues, a feature of F2 (Future 2) curriculum thinking. This is modelled in Fig. 6.1 as an arrow linking curriculum making directly to agency. This would be akin to teachers telling their students what to think and how to behave. This would be encouraging a particular agency, a set way of behaving that is not a free choice of the educated pupil; a teacher telling their pupils to buy Fairtrade products or to vote in a particular way in an election would be an example of not enabling the pupils to take on agency, but is 'morally careless' teaching.

Capabilities across a School: A Curriculum 'Big Picture'

The concept of capability can be used as a way for teachers and school leaders to think about the purpose and structure of a whole school curriculum. It links curricular aims through a knowledge-led, subject-based curriculum to pupil outcomes. It recognises that each school subject, such as geography, enables a significant part of a much broader educational outcome, articulated as knowledge-based capability, to develop. As a way of thinking it is an alternative to the much critiqued National Curriculum 'big picture' (QCA 2008) and Habits of Mind (HoM, e.g. Boyes and Watts 2009) discussed in Chap. 1, which can also attempt to provide a unifying concept cutting across subjects. Both these can be considered expressions of Future 2 curriculum thinking.

A capabilities perspective can be used as a means for schools to be explicit about the value of a knowledge-led curriculum. Schools can claim proudly that an aim of education in their school is to introduce young people to the best thoughts and ideas that humanity has ever discovered rather than trying to express it implicitly through examination outputs. The focus of teachers across a school on teaching powerful knowledge ensures pupils have a grounding in the nature of subjects and are not just learning facts to pass an exam.

Using the thinking offered by the capabilities approach is one possible way of achieving a Future 3 curriculum in a school. Creating a Future 3 curriculum across a whole school is conceptually very appealing, but much more research is required to see if it is actually possible. The challenge has been recognised in part by Tim Oates (2018) who identified:

> If Future 1 and Future 3 appear to have 'space' between them, then this could appear like the distance between the Earth and Moon. But if this is the analogy, then the distance between 1 and 3 is nothing compared to the difference between these and Future 2—which in epistemological terms is in a galaxy far, far away. While Future 1 and Future 3 may require a short and intensive debate to resolve the practicalities of translation into legitimate curriculum policy, Future 2 was embedded in an entirely different and outdated conception of 'knowledge'. (Oates 2018, p. 159)

For Oates, recognising a Future 2 curriculum and how it is different from a knowledge-led curriculum in a school is conceptually easy. The challenge lies in separating an F1 curriculum from an F3 curriculum as both are centred on knowledge through academic subjects. They are separated conceptually by the type of knowledge being taught, and the way that knowledge is recontextualised for young people. As Lambert (2018) discusses:

> It is because of the nuanced variance between Futures 1 and 3, together with the strong gravitational pull exerted by the familiarity of Future 1, that careful and regular thought needs to be given to the question of how to achieve the *epistemic quality* desired in Future 3—and how to make this accessible to students of all backgrounds and circumstances. (p. 9, original emphasis)

To achieve an F3 curriculum across a whole school, without defaulting to the 'familiarity' of an F1 curriculum, is a challenge. Capabilities think-

6 The Potential of a Future 3 'Capabilities' Curriculum

ing can be of use in this regard. To achieve the sort of thinking required involves a whole school approach, getting teachers to work in teams of subject specialists to identify the ways in which their subject knowledge is powerful for young people. This could be conducted during an in service teacher training session; this can then be used as the basis for rigorous curriculum planning, possibly allowing teachers to use a version of the practical GeoCapabilities Framework table (Fig. 5.2) to assess knowledge. It is the focusing of teaching on this and not solely on examination preparation, through the awareness and recognition of varying ideologies, that can then enable the various capabilities discussed previously.

Work to develop whole school thinking through this approach is in its infancy; in Standish and Cuthbert (2018), experts from a variety of subjects were invited to explore some of these ideas within their own subject expertise, such as mathematics (Crisan 2018), languages (Lawes 2018), physics (Sturdy 2018) and art (Powell 2018). Work with groups of teachers in schools has also started to explore some of these ideas as part of ongoing research into whole school capabilities. Table 6.1 shows some initial results from two sets of subject specialist school teachers exploring the powerful knowledge of their subjects as part of separate whole school teacher training exercises. These schools are both UK Secondary schools and the subjects highlighted here are religious studies, drama and classical languages.

The teachers' work here begins to help explore how different subjects might provide powerful knowledge for young people. In the case of religious studies, there is consistency across the two sets of teachers in the schools with the idea that the subject is able to offer factual knowledge of different religions as well as nurturing a sense of debate and discussion of philosophical issues. The two sets of drama teachers are consistent in the idea of the importance of practical performance and the theoretical aspects of exploring dramatic text but there are differences; the teachers in school A are the only ones who consider the reaction of an audience as being a key feature of the knowledge of drama education. Teachers of classical language also agree with the idea that their subject provides a basis for an understanding of more modern language traditions and cultures but only one of the two sets of teachers considers the significance of 'translation' as a key part of the powerful knowledge of classical language.

Table 6.1 School teachers exploring the powerful knowledge of their subjects

Subjects	School A	School B
Religious Studies (including philosophy and ethics)	Rigorous analysis of concepts and arguments and coming to conclusions. Factual and descriptive knowledge of alternative schools of thought/faiths and contexts. Putting together concepts and ideas to compare, contrast, argue and debate.	Increasing factual knowledge of the views of major world religions on the BIG metaphysical and ethical questions. Nurturing pupils' own viewpoints on these issues and fostering their spiritual awareness through challenging viewpoints.
Drama	Theoretical and practical procedures that explore texts and how they are seen and reflected through a performative lens. Critically reading our world in a way that transcends text and vocal communication that can both connect and polarise audiences. Developing dramatic literacies and social skills.	Knowledge of social, cultural, historical and political aspects of theatre and the human condition. Theoretical and practical application of key concepts in historical and contemporary theatre. The ability to engage, explore, empathise and challenge perceptions of the human condition.
Classical languages (Latin and Greek)	Deep cultural and philosophical insight into the origin of Indo-European human thought.Cognitive development and focus of logic, decoding, creativity and justification of translational methodologies.	Understanding of the languages which form the grammatical and linguistic basis of both English and many other languages.Reading, in their original text, works of literature which have provided the model for influential works throughout history.Opportunity to apprehend the entirety of a culture in its linguistic, historical, philosophical and aesthetic legacy.Recognising and evaluating the continuation and development of classical ideals and theories throughout subsequent cultures and civilisations.

Many schools in the UK do not offer courses in classical languages; these are often seen as the preserve of the more elite independent schools. The idea is a powerful classical language knowledge could be used as a means to justify the inclusion of classical languages in a whole school curriculum. The framework of thinking that powerful knowledge and capabilities provide can be used to express the potential of different subjects in a curriculum.

The ideas presented in Table 6.1 are a first attempt to move these ideas beyond geography education and to work with teachers in schools to develop the notion of powerful disciplinary knowledge across a school curriculum. Much more work is needed, across a broader range of subject specialists and in a wider range of schools but the ideas shown here can be seen as a first attempt at this process.

GeoCapabilities Curriculum Thinking in Action

The capabilities approach to curriculum thinking can be used to help geography teachers think deeply about what they choose to teach and why. It can help them to ensure a powerful geographical knowledge dimension is included in their lessons and enable them to clarify pupils understanding on issues rather than indoctrinating pupils to a specific response. A few examples of how this thinking can benefit teachers have been alluded to already in this book: Standish's explanation of teaching 'Fairtrade'; the vignette of powerful knowledge about coastal erosion; and the curriculum making workshop in the doctoral research about 'Russia'. This section outlines how capabilities thinking led to two of the lessons in *What's the Use: How can Earth meet our resource needs?* (Bustin 2015), a resource book for teachers with pre-planned lessons on 'Natural resources', another of the topics on the 2013 geography National Curriculum.

The whole resource book itself is based around an enquiry approach, with questions framing each of the lessons; the structure of the ten-lesson key stage 3 course focuses on essential resources of energy and food, and the non-essential resource of diamond. The two lessons that focus on food, 'something fishy going on' and 'a net result', focus on tuna in the Indian Ocean. Table 6.2 shows the outline of these two lessons.

Table 6.2 Two three part lessons from the medium term plan of a sequence on tuna fishing, part of a larger topic on natural resources (from Bustin 2015)

Lesson	Key questions	Learning objectives	Teaching and learning	Assessment opportunities
Something fishy going on	What type of resource is tuna? How does tuna fit into the marine ecosystem?	To classify tuna as a resource. To understand how tuna fits into the marine food web and how it can be affected by changes to the ecosystem. To consider the ownership of ocean resources and how they should be shared.	**Starter:** Whole class discussion: is tuna a renewable or non-renewable resource? **Main:** Students compare food web interpretations and answer questions. They map Indian Ocean countries and answer questions. **Plenary:** Whole class discussion: who owns the fish in the sea?	Food web questions. Accuracy of map and question answering.

(continued)

6 The Potential of a Future 3 'Capabilities' Curriculum

Table 6.2 (continued)

Lesson	Key questions	Learning objectives	Teaching and learning	Assessment opportunities
A Net result	How sustainable is tuna as a food resource? Should tuna fishing be banned?	To understand what makes fishing sustainable. To acknowledge that are a range of opinions about the tuna industry. To consider their own response to the issue of unsustainable tuna fishing.	**Starter:** Students are introduced to the concept of overfishing. **Main:** Students watch a video clip, or undertake internet research, and answer questions. They undertake a talking heads exercise to assess whether tuna fishing should be banned. They compose an email to one of the characters from the talking heads exercise, supporting or arguing against their opinion. **Plenary and homework:** Students decide for themselves if tuna fishing should be banned and hold a class vote. They investigate fish farming as an answer to the problem of overfishing.	

The structure of these two lessons is based on the three-part lesson with starter, main and plenary activities. It is designed to be used with key stage 3 pupils (11 to 14 years old) and there is probably too much for two lessons here but that was partly the intention; it gives teachers choice in what aspect they want to focus on. Suggested teaching and learning activities are given here and the book itself comes with all the associated resources needed to run the lessons as presented in the book. Teachers are encouraged to amend the lessons as they see fit, hence the title of the series, 'teacher toolkit', with the activities and resources acting like tools that teachers can use depending on their setting. In many ways that freedom to adapt the plans fits in with the notion of 'structural features'; teachers can use these lessons in any order they wish, they can use some or all of them and this might depend on their school ethos, their own values about geography teaching or other areas of the school curriculum.

The practical planning tool of the GeoCapabilities Framework table explained in the previous chapter (from Bustin et al. 2017, Fig. 5.2) was used to identify the powerful knowledge component of the sequence, and Table 6.3 shows how the knowledge taught in these two lessons can fit into the three expressions of powerful geographical knowledge. This ensured there was a geographical knowledge component to the sequence.

Table 6.3 shows that activities for these lessons can fit into all three of the expressions of the powerful knowledge of geography, and in fact the three are sequential in this case. Firstly, the pupils map and use data to

Table 6.3 The use of the practical GeoCapabilities Framework to map out the geographical knowledge component of the two-lesson sequence on tuna fishing

Deep descriptive 'world knowledge'	Theoretically informed relational understanding of people and places in the world.	Propensity and disposition to think about alternative social, economic and environmental futures.
Mapping the Indian Ocean and surrounding countries. Demographic and economic data about those countries to determine levels of development.	Ecosystems and food webs—how tuna fits into its ecosystem and impacts of overfishing. Globalisation of tuna trade.	Sustainability of tuna resources and choice about consumption patterns. Understanding of key players and their attitudes.

explore countries around the Indian Ocean. This would in part help explain why fishermen in those nations would be keen to catch and sell lucrative tuna as a resource. The next part of the lesson looks at how tuna fits into its food web. The pupils engage with this by taking a species out of the web and identifying the impacts of this on the other species in the food web; not every species will be affected if some are removed. This sort of activity could be used in a Biology lesson, but in this context it enables the pupils to think geographically, as it is here that pupils can look at what happens if too many tuna are removed before nature can replenish them, and what happens with the sea turtle and dolphin populations which often get caught up in the fishing nets. This knowledge is significant for the final stage of the lesson sequence. Here, the pupils are engaging directly with values. The large trawler boats enable countries to fish on a large scale and bring in much needed money for their various economies, but with the unwanted 'bycatch' of turtles and dolphins, and the potential for overfishing this could be seen as unsustainable. By contrast, small scale fishing of tuna caught by 'pole and line' methods are less financially beneficial but more sustainable. The final enquiry question, 'should tuna fishing be banned?', enables pupils to explore all these issues and to come to their own response to the question.

Many leading supermarkets are now keen to show they are supporting pole and line caught tuna, but trawler caught tuna is still available in shops and is often cheaper than its more sustainable counterpart. This means there is a strong values dimension to the work, which might suggest a 'radical' ideological perspective, yet the capabilities approach cautions against the 'morally careless' approach of values transmission. Pupils use knowledge of economic data, knowledge of food webs and knowledge of fishing methods built up in the previous lesson activities to clarify their position in relation to sustainable fishing.

The capabilities to which these lessons contribute enable the pupils to be more considerate when making their own consumption choices. When they are in the supermarkets buying tuna the knowledge they have gained will enable them to make an informed choice about whether they buy pole and line caught tuna or not. Morally careless teaching would tell them to do so; capabilities gives them the knowledge and the choice. Within the broader framework of the rest of the lessons in the whole sequence, pupils can use their knowledge to make informed choices about their relationship to natural resources around the world.

The capabilities approach enables this depth of curriculum thinking and this can invest geography education in schools with its vast potential. Translated across all subjects it has the possibility of enabling a Future 3 capabilities curriculum for a whole school.

Challenging the Conceptual Approach

Notions of powerful knowledge, the F3 curriculum and GeoCapabilities have been critiqued by a number of writers.

Powerful Knowledge

Of all the concepts explored in this book, Michael Young's 'powerful knowledge' has become the most celebrated, particularly given its prominence in discussions around the 'knowledge turn' of the National Curriculum. One critique of the nature of powerful knowledge came from Dr Mary Boustead, the influential joint head of the National Education Union. For her, a curriculum based around powerful knowledge perpetuates a version of knowledge that downplays the significance of the contributions made to knowledge by writers from a range of ethnic and cultural traditions. As she argued:

> If a powerful knowledge curriculum means recreating the best that has been thought by dead, white men—then I'm not very interested in it. (Ward 2018)

She illustrated this by referencing her time as an English teacher. Whilst she admitted she had no problems teaching Shakespeare, she was always keen to ensure the children in her multi-cultural classroom were exposed to writers from a range of traditions.

It is easy to dismiss her claims as simple misunderstanding of powerful knowledge; she is identifying here what Young (2008) himself identified

6 The Potential of a Future 3 'Capabilities' Curriculum

as knowledge 'of the powerful', with the powerful here being 'dead white men'. Young (2008) identified the inadequacies of this thinking himself and moved on to 'powerful knowledge' as a response to this thinking. The power of epistemic communities working together to create knowledge is the basis of the social realism ideas presented in previous chapters. The response to the simple dismissal of powerful knowledge like this is a return to F2 thinking where the aims of an English Literature education (in this instance) is more about ensuring some sort of author diversity, rather than helping pupils explore the value of the 'powerful knowledge' of English Literature as a subject. Her narrow view of knowledge is related to the F1 curricular vision of lists of facts to be learnt that she fears bear no relevance to the lives of her pupils.

Despite the response to her critique, there is still value in her criticism. If powerful knowledge is created by epistemic communities of experts then the extent to which these communities are drawing on culturally diverse knowledge, and the appropriateness of this, is questionable. If subjects have developed over the past 150 years, then key thinkers from the past who are now considered to have played a major role in the development of a discipline may well have been men and may well have been white. This is not a reason, however, to dismiss subject knowledge completely. If the gender and ethnicity of those creating knowledge was significant, 'powerful knowledge' would recognise these potential one-sided versions of knowledge and engage in debate about this.

Boustead's remarks also refer to the extent to which her multi-ethnic pupils felt engaged by knowledge created by 'dead white men'. A key factor in curriculum making, and modelled as part of the structural features of the capabilities approach, is an understanding of the pupils in the classroom. A teacher who knows their students will be able to choose knowledge and content that have the chance to engage and inspire. If they have a clear understanding of what makes their subject knowledge powerful then they are able to produce a relevant curriculum that is also empowering.

Margaret Roberts (2013, 2014) has been critical of the notion of 'powerful knowledge' as being separate from 'everyday' knowledge that children bring to the classroom each day. Roberts (2014) cites the work of Vygotsky (1962), who identifies close links between spontaneous or 'everyday' concepts and scientific concepts. The former develops out of a child's interactions with the world and the latter have a more formal knowledge structure. Vygotsky's work identifies the close connection between these two ideals and Roberts (2014) agrees, arguing that for the subject of geography the 'everyday' understanding of the world is an important resource for teachers to draw upon in the classroom, which should therefore be considered 'powerful'. Many important curriculum development projects led by the Geographical Association (such as the 'Young Peoples Geography' Project 2009) give support to teachers to help them draw out this experience from students in the classroom. She even points out that many avenues of research in academic geography focus on the ways in which young people construct their own meaning from their everyday interactions with places (e.g. Skelton and Valentine 1997). This gives academic validity to the sorts of geographies that children develop. It is for this reason she takes issue with the idea of 'powerful knowledge' being distinct from everyday knowledge. As she argues,

> the issue for me is not whether the school geography curriculum should develop students' knowledge beyond their existing experience. I know of no geography curriculum for any age group based entirely on students' everyday knowledge. The issue is about whether a geography curriculum document setting out what students are required to study, should exclude, as Young suggests, students' everyday knowledge. (p. 193)

Margaret Roberts and Michael Young took part in a debate of these issues in front of an audience of geography educationists in May 2013; both talks are available online (Young 2013; Roberts 2013). Yet their positions are not as distinct as previously thought. The separation of everyday and powerful knowledge that Michael Young has been so careful to distinguish between suits the knowledge developed through scientific methodologies. Children might well know that plants grow, but to understand how plants grow requires a specific set of ideas which can be considered the 'powerful knowledge'. Children will not be able to gain this knowledge without teachers. For school geography, the everyday experience of children is an important point for teachers to start teaching a topic but it is only a starting point; from there larger concepts and knowledge structured can be added, as Roberts herself acknowledges (2014). All children will come to a classroom with an experience of a city, or a river but they do not come with an understanding about how the city develops and changes, nor how helicoidal flow in a meander bend shapes the river landscape. It is not the everyday knowledge that is powerful, but how skilled teachers help students to make sense of their experience that gives it power. This nuanced distinction is key and emphasises again the role of the skilled teacher.

The debate about what constitutes powerful knowledge continues, particularly ideas about the place of faith and religious knowledge in the classroom or the 'power' of indigenous knowledges developed within non-disciplinary communities. Even the word 'powerful' has been critiqued as having masculinist overtones, which could be considered unhelpful. Research into the term, its basis and ideas will continue to hone and shape the concept.

The Challenge of a Subject-Based Curriculum

Much has been written already in this book about the differences in opinions between those advocating a skills-based and others advocating a knowledge-led, subject-based curriculum. Yet a more specific critique has been levelled at the official National Curriculum Review document (DfE 2011) which led to the 'knowledge turn' which seemed to be based around "a few privileged subjects" (Brown and White 2012). As Brown and White (2012) argue:

> Citizenship is relegated… the Review doubts that, along with Design and Technology and ICT, it has 'sufficient disciplinary coherence' to be a discrete and separate National Curriculum 'subject'… why is Modern Foreign Languages (MFL) retained as a National Curriculum subject? What is its 'distinct way of investigating'? Is it in the business of investigating at all? Where are its 'theories'? (Similar points could be made about PE). Science seems to be the paradigm, on this definition, for a pukka National Curriculum subject. Even geography is in choppy water.

Their argument is about the disciplinary coherence of certain subjects, and they argue that science and maths have more rules, norms and knowledge coherence than history, geography or MFL. This could be related to the ideas of knowledge structures presented earlier; for the sciences and maths, the 'verticality' (as Bernstein 2000 would describe) provides an obvious curricular coherence. The more horizontally structured subjects seem to be less coherent and as such could be challenged as a form of powerful knowledge. In Chap. 3 it was recognised that geography has classical origins, yet the knowledge of the world that was being generated was not called 'geography' at the time, so there is not a long heritage of disciplinary coherence, which challenges the ability to create powerful knowledge (this notion is explored thoroughly in Standish and Cuthbert 2018). This is in part where Martin Robinson's (2013) work on the classical trivium is of relevance. It is the organisation of knowledge through this structure that has a longer heritage, although this structure is not mirrored in contemporary discourse, so there is no means to generate 'powerful knowledge'.

Capabilities, built around powerful knowledge, could be a means to address this challenge. The discussion around which subjects have a claim to disciplinary uniqueness, and therefore can be 'powerful' for young people and part of a Future 3 capabilities curriculum, and which subjects are not is still a challenge. Young (2012) offers some clarity by differentiating between disciplines and school subjects:

> Subjects … are different from disciplines. They are not a source of new knowledge; they are different from disciplines but draw on disciplinary concepts and organize, sequence and select from them in ways that have proved most reliable pedagogically.

It is subjects, not disciplines, which form the basis of a school curriculum. The ideas around recontextualisation discussed earlier provide a mechanism for the ways skilled teachers can 'draw' on the concepts from the disciplines. Yet this does not provide an adequate argument for which subjects should be part of a curriculum, other than historical tradition. As Young (2012) continues:

6 The Potential of a Future 3 'Capabilities' Curriculum 179

Subjects and disciplines have a long history, taking us back 150 years and earlier... The relative stability of subjects and their boundaries is partly why parents trust schools and partly why employers invariably prefer subject-based (or academic) to vocational qualifications when recruiting new staff.

The idea of stability and rigid boundaries can be a useful starting point about which subjects do and do not have a claim to be a powerful knowledge. Despite the challenges in expressing this for geography, this book has shown how writers such as Lambert (2018), Lambert and Morgan (2010) and Maude (2016) have expressed the powerful knowledge of geography. The next main avenue of research, if the idea of powerful knowledge and the F3 curriculum are a useful conception, is for other subject disciplines to identify what makes their subjects uniquely powerful for young people, and therefore justified as part of a knowledge-led curriculum. This work has already begun, as discussed previously.

Yet there remains a challenge. Many schools, particularly those independent of the National Curriculum, teach classical languages such as Latin and Classical Greek, and most schools now teach Computing, which has seen numbers of pupils studying the subject at GCSE (General Certificate of Secondary Education) increasing (BBC News, Wakefield 2017). If these subjects can be found to be uniquely 'powerful' in some ways, then their inclusion in a knowledge-led curriculum can be justified and this then produces an argument for their inclusion in all schools. This can still result in an overcrowded curriculum, with pressure on curriculum time not coming from generic skills but from other school subjects.

The thinking offered by the capabilities approach does have a response to this, at least in part. The knowledge-based capabilities that pupils would develop through their schooling will differ within and between schools. A school that offers Latin and Greek will enable their students to develop knowledge of those subjects which will contribute to those students' overall capability set. These capabilities would not be available to students in schools where these subjects are not on offer. Within a school, a pupil's capability set is a product of the subjects they choose to study. Across a whole school, the curriculum on offer to pupils is a product of the thinking done by school leaders. Headteachers and those responsible for the curriculum are the ones who decide what subjects are deemed 'powerful' enough for inclusion in a school curriculum and how much curricular time is given over to each subject. These are not easy decisions to make and get to the heart of why the issues discussed in this book are important so school leaders can make informed curricular decisions.

Capabilities: An Unworkable Framework?

During the doctoral research, the teachers in the workshop ignored the influence of the Habits of Mind (HoM), a major overarching curriculum structure at the case study school, when curriculum making. The capabilities approach could be ignored in a similar vein by teachers. This would render the concept a theory with little practical use. Both the HoM and the capability approach are curricular concepts that have similar traits. Both are focussed on pupils, identifying how children can think with the education received, either through the organised disciplinary ways of thinking as offered through the capability approach or through broad scale competencies of the HoM. These ways of thinking provide an aim for a curriculum. Yet the capability approach to curriculum thinking would not be subverted or ignored in the way that the HoM has been. Capability is fundamentally different to HoM. To incorporate HoM into curricula, teachers actively need to alter the content of what they are teaching to incorporate these ways of thinking. This means changing the knowledge basis of their lessons, splicing extra content to cover a particular habit which takes them away from the subject knowledge. In geography, a teacher preparing to teach a lesson on Russia might feel the need to include something to help the children to "think flexibly" (one of the Habits). Rather than being something to reflect on after a sequence of lessons, the HoM have been interpreted too literally by some teachers, and this additional 'pressure' to incorporate these cross-curricular competencies has led to teacher resentment, particularly as many teachers will feel they already incorporate these elements into their teaching. There is also a fundamental message about the nature of subject knowledge that the HoM gives; it suggests that subject knowledge is simply the means to develop the HoM. It suggests that the habits are the more significant curricular idea, more important than subjects which are therefore a distraction.

The capability approach to curriculum thinking does not require teachers to alter their curriculum content. In fact, it raises the importance of subject knowledge as the most important idea in curriculum. The idea of powerful subject knowledge is much more tangible than the HoM, and easier for teachers to see as relevant. Teachers are still, mostly, subject specialists and capability enhances their professionalism and values their knowledge basis. This is empowering for teachers, and teachers should feel encouraged by the thinking that capabilities offers. Capability is not an idea that is simply tacked on to an existing curriculum structure, unlike the HoM, but is integrated within a subject-based curriculum. This means capability is a much more significant and robust way of thinking about the secondary school curriculum than the HoM.

GeoCapabilities: Michael Young's Critique

Despite using much of Michael Young's work (e.g. 2008) in this book, Young (2011) was critical of Lambert's early articulations of GeoCapability, arguing:

> [A]s a curriculum principle it is too general to underpin the crucial role of schools in transmitting the 'powerful knowledge' on which a student's future 'capability' will depend. (p. 182)

For Young (2011), the ideals of social realism express the significance of knowledge in schools and thus capabilities becomes a framework that is too general. This critique provides an opportunity to open a discussion about the perceived mismatch between the central concepts of social realism and the capabilities approach. These concepts are both central to this book, and it is the difficulty in reconciling these different approaches which may be behind Young's (2011) views.

The notions of social realism from Young (2008) and GeoCapabilities (e.g. Lambert and Morgan 2010) have similar ideals. Both are attempts to ensure knowledge is a central consideration of educational discussions.

The type of knowledge to which they both aspire is also similar, created by epistemic communities of disciplinary experts, which makes it 'better' knowledge. Both approaches are therefore also clear in the role of subject disciplines; better knowledge is created and maintained within subject groupings. Both concepts also have people at the centre of the approaches. In social realism, these people are the knowledge creating subject experts, and their knowledge is thus socially constructed. The capability approach is a concept for teachers working in schools which emphasises their role as subject experts. Thus, both approaches apply to educated people who have a relationship to a wider set of disciplinary norms.

Yet social realism and the capability approach are fundamentally different concepts. Neither social realism nor the capability approach was initially designed to be applied directly to a school curriculum. Writers have used the concepts and ideas and applied them to discussions in schools. Social realism traces its roots to the sociology of education whereas the capability approach stems from welfare economics and the discourse on human rights. Thus, the capability approach has been applied to issues of education rather than emerging naturally from it. The principle difference between the two concepts goes back to the nature of knowledge. Social realism provides a set of principles under which knowledge can be created. When translated into the school curriculum, this becomes 'powerful knowledge', and this approach therefore focuses on the inputs to a curriculum. Social realism is not concerned with educational outcomes or overarching curricular aims; it is simply about the quality of the knowledge that is the basis for a curriculum. In many ways, social realism thus articulates a narrow focus on a key area of curriculum discourse. The capability approach, conversely, is a much larger scale framework that encompasses curricular aims as well as outcomes. It is concerned as much with *what* is being learnt (the only consideration of powerful knowledge) as to *why* this is being learnt. Capability focuses on the outcomes of a curriculum.

The difference between the focussed nature of social realist knowledge as a basis for the curriculum and the broader framework of the capability approach may help explain Young's (2011) critique of capability being 'too general'. Yet part of Young's critique also stems from a lack of understanding of the principles. He was responding to Lambert (2011), in which the ideas of the capability approach to education and GeoCapability specifically were in their infancy and perhaps at that stage were 'too general'. No empirical basis had been provided at this point, and the international projects were only starting. Thus, Young was responding to a concept that was underdeveloped in the academic literature.

Since his 2011 challenge, Young has remained interested in the notion of the capability approach to education. In a more recent email exchange with David Lambert, he admits a lack of understanding of the initial concept. When referring to the work of Nussbaum (2000) with regard to the initial capability approach, he admits:

> It (the capability approach) has two distinct features which I did not realise- it is a moral/political theory and a normative not an explanatory theory ... It prescribes, based on principles, but does not set out to explain the why or the how; this makes it very different from sociology and not directly comparable. And this helps me because at least I (now) know what she is trying to do. (Michael Young, 18 April 2015, personal communication[1])

The notion of capabilities being a normative theory is the idea that it sets out a set of principles that should be 'normal' practice. For knowledge-based capability, these principles are based on powerful knowledge being the basis of a subject-based curriculum and therefore foreground the important role of teachers in creating and 'making' the curriculum.

GeoCapability is not too general a theory; it articulates a means by which powerful knowledge can be embedded in a curriculum by ensuring a subject-based, knowledge-led curriculum. The capability approach relies on teachers making decisions about what, why and how to teach with a shared understanding of the importance of subject powerful knowledge and pupil outcomes. It is hoped that by basing discussion on an empirical set of data, and developing a robust discussion about the nature and possibilities of the capability approach, and how it is different to social realism, some of the concerns of Young (2011) have been addressed.

---

[1] Email communication made available by kind permission.

These critiques are all part of an ongoing discussion about the nature and validity of the central concepts in this book. There is more research that needs to be conducted into all of these ideas, as outlined in Chap. 5.

Meeting the Potential: Is an F3 'Capabilities' Curriculum Possible?

This book has presented a vision for a school curriculum—backed up by research—that has the potential to offer young people access to world-class knowledge that is empowering, knowledge that develops capabilities that enables young people to make choices when they reach their adult lives. Many teachers and some whole schools already exhibit aspects of F3 thinking, and the workshops in the research provided an opportunity to explore this. But if F3 curriculum thinking, built on the powerful knowledge of subjects, is to be a useful construct in all schools there are some significant implications that this would entail which might challenge the way the curriculum is currently conceived and structured.

1. A Powerful Knowledge-Led Subject-Based Curriculum

The obvious first step towards a capabilities curriculum is to have a subject-based curriculum. To make it 'powerful knowledge' led, subject teachers need to identify what makes their subject powerful for young people and to be able to express this. This then informs curriculum making. Teachers actively ensure powerful knowledge is at the heart of every lesson, using their pedagogical abilities to help pupils engage with the materials to help develop their knowledge led capabilities. This is not to say that skills are unimportant in a school. Many skills develop naturally through a subject-based curriculum, and can be reflected upon by pupils (such as Habits of Mind). 'Extracurricular' activities, whole school sports programmes and outdoor adventurous activities form important parts of school life, often running outside the timetabled lessons, at lunch times and after schools, and these can bring benefits to young people. But the principle is that these occur alongside the timetabled school lessons, not at the expense of knowledge and subjects.

2. A Focus on Outcomes Not Outputs

To achieve the sort of grand vision to which capabilities aspires, linking the aims of education, through powerful subject knowledge, to the results of education, school leaders, teachers, pupils and parents need to focus on the holistic 'outcomes' of education, and not simply on the 'outputs' of exam grades. This is a significant challenge in a system that ranks schools on the basis of examination results and publishes these annually. If the reputation of a school lies solely in examination results it creates a system where teachers feel heavily 'controlled' by examination criteria, pressure is felt to teach to the test and pupils and parents feel under pressure to succeed within these criteria. As a society we have learnt to 'value what we can measure', rather than finding a means to measure what we truly value (a sentiment from Hargreaves et al. 2014). Teachers need to prepare students for examinations, and this produces one form of data that can be of use, but it is overemphasis and control has created an exam focussed school system.

Capabilities cannot be measured and placed in a league table. They represent intellectual freedoms and ways of thinking. This makes it a difficult aim of school to 'sell' in an environment which seems to encourage competition between schools. Yet the stifling influence of examination success could be preventing schools, and the young people they educate, from really achieving their full academic potential.

3. A Focus on 'Curriculum' as Well as Pedagogy

Given the controlling influence of the National Curriculum of the past which detailed specific knowledge to teach, many teachers have not had to engage with thinking about what is being taught. Most innovation, most books and ideas focus on innovative ways to teach content, rather than conceptual discussions about what is being taught. Leading textbook series exist for all levels and all subjects, giving a legitimacy to a set of content and removing the need for teachers to think about curriculum for themselves. The capabilities approach requires teachers to be the curriculum leaders and to use their understanding of their subjects to choose the best knowledge to teach. This means teachers need to spend time really focussing on what they are teaching and why. Once this is established, time can then be spent on pedagogical discussions about how to teach that content.

4. A Coherent Curriculum

The capabilities approach provides a means to provide a coherent picture of the school experience for young people. It enables them to understand what connects their seemingly unlinked subjects together into a bigger picture of their education. This coherence can be of use to articulate the aims of a school and how knowledge is at the heart of every school's aims. This coherence also aids subject teachers, with the powerful knowledge proving a means to ensure that young people are introduced to the disciplinary thoughts of the subject allowing them 'epistemic access' to the discipline and the challenges this creates.

5. A Subject Specialist in Front of Every Class

The level of curriculum thinking required to teach powerful knowledge, and to think about why that knowledge is being taught, and how it links to broader aims requires specialist thinking. It requires teachers to be not only subject specialists but also teaching experts who understand the significance of careful curriculum making and recontextualisation. It is subject specialists who can provide the awe and wonder of a subject, and only they who can induct young people into a 'discipline'. The importance placed on teachers to enable a Future 3 curriculum is the reason the GeoCapabilities 2 project was subtitled 'teachers as curriculum leaders'. It is the leadership aspect of the teacher role which enables them to make choices during curriculum making.

Ensuring a fully qualified, subject specialist teacher in front of every class is challenging. Wider challenges of teacher recruitment and retention detailed in earlier chapters cannot be solved by one school, but lowering entry criteria for potential teachers and finding cheaper, short cut training programmes, often school based, are not the solutions. Universities are key to the process of high-quality teacher education.

6. Teachers As Professionals

It is the teachers who enable the curriculum in the classroom for the pupils they teach. The aspirations of the capabilities approach give the power and control to teachers and not to examination criteria, or other curricular needs. Teachers must be given freedom to design and implement exciting and engaging lessons. To achieve this, teachers need to be fully trained subject specialists and trusted by leadership teams to enact an enticing curriculum. Yet many teachers do not have the level of confidence in their own abilities, nor the experience, to think creatively about the curriculum.

A possible model to aspire to can be demonstrated in Finland, where teaching is a highly competitive profession, for highly qualified graduates. As Hargreaves et al. (2014) explain:

> All teachers possess Masters Degrees; the acceptance rate into teacher education programmes stands at less than 10 percent; and the inspiring dream that assigns national importance to education and educators draws the brightest and most committed graduates into teaching. (p. 146)

If this model were to be adopted elsewhere, teachers would be respected across society and seen as 'public intellectuals', enacting and making engaging curricula for the young people of the nation. It seems their status is much more 'technocratic servant', teaching to the test, getting pupils through examinations and focusing all their energies on behaviour management, lesson timing and pacing. It is this grand role for teachers that can enable teaching of the highest quality. A consequence of this is, of course, a need to ensure teachers are well remunerated for their efforts, and the lifestyle of the job is conducive to the level of thinking required.

Conclusions

The capabilities approach can be one way in which the aspirations of a Future 3 curriculum could be met. There are significant challenges which need to be met in order for the sorts of grand visioning presented here to be enabled across a school or a country. What the capabilities approach does enable is a means for teachers to realise the academic potential of their subjects. By focussing on the powerful knowledge of their subjects, young people from all backgrounds can develop capabilities to think about the world and make positive choices about how to live. This has the potential to envisage a world class education.

> **Questions to Consider:**
> 1. Can you relate the model of capabilities curriculum thinking to your own school setting? Can you identify the various curriculum pressures and the extent to which they enable or constrain teacher choice?
> 2. To what extent can capabilities provide a coherent vision for a whole school curriculum?
> 3. Think of a sequence of lessons you teach/have taught. Would a GeoCapabilities perspective on this sequence offer a new perspective? In what ways?
> 4. How do you respond to Michael Young's critique of GeoCapabilities?
> 5. Which of the challenges presented here resonate mostly with you? What could be done to overcome some of the challenges?

References

Bernstein, B. (2000). *Pedagogy, Symbolic Control and Identity: Theory, Research and Critique* (Rev. ed.). London: Taylor and Francis.

Boyes, K., & Watts, G. (2009). *Developing Habits of Mind in Secondary Schools*. Moorabbin, Australia: Hawker Brownlow.

Brown, M., & White, J. (2012). An Unstable Framework—Critical Perspectives on the Framework for the National Curriculum. Retrieved December 2018, from https://www.newvisionsforeducation.org.uk/2012/04/05/an-unstable-framework/

Bustin, R. (2015). *What's the Use: How Can Earth Meet Its Resource Need? Geography Key Stage 3 Teacher Toolkit.* Sheffield: Geographical Association.

Bustin, R., Butler, K., & Hawley, D. (2017). GeoCapabilities: Teachers as Curriculum Leaders. *Teaching Geography, 42*(1), 18–22.

Crisan, C. (2018). Mathematics. In A. Standish & A. S. Cuthbert (Eds.), *What Should Schools Teach? Disciplines, Subjects and the Pursuit of Truth.* London: UCL IOE Press.

DfE. (2011). *Review of the National Curriculum in England.* London: DfE.

Giddens, A. (1984). *The Constitution of Society: Outline of the Theory of Structuration.* Oxford: Polity Press.

Hargreaves, A., Boyle, A., & Harris, A. (2014). *Uplifting Leadership: How Organisations, Teams, and Communities Raise Performance.* San Francisco, CA: Jossey-Bass.

Lambert, D. (2011). Reframing School Geography: A Capability Approach. In G. Butt (Ed.), *Geography, Education and the Future.* London: Continuum.

Lambert, D. (2018). *Teacher Preparation, Curriculum Leadership and a Research Engaged Profession: The Case of Geography.* Working paper produced for the Geography Education Research Collective (GEReCo).

Lambert, D., & Morgan, J. (2010). *Teaching Geography 11–18: A Conceptual Approach.* Maidenhead: OUP.

Lawes, S. (2018). Foreign Languages. In A. Standish & A. S. Cuthbert (Eds.), *What Should Schools Teach? Disciplines, Subjects and the Pursuit of Truth.* London: UCL IOE Press.

Maude, A. (2016). What Might Powerful Geographical Knowledge Look Like? *Geography, 101*(1), 70–76.

Nussbaum, M. (2000). *Women and Human Development: The Capabilities Approach.* Cambridge: Cambridge University Press.

Oates, T. (2018). Powerful Knowledge—Moving us All Forwards or Backwards? In D. Guile, D. Lambert, & M. Reiss (Eds.), *Sociology, Curriculum Studies and Professional Knowledge: New Perspectives on the Work of Michael Young* (pp. 157–168). Abingdon: Routledge.

Powell, D. (2018). Art. In A. Standish & A. S. Cuthbert (Eds.), *What Should Schools Teach? Disciplines, Subjects and the Pursuit of Truth.* London: UCL IOE Press.

Qualifications and Curriculum Authority (QCA). (2008). *The Big Picture of the Curriculum.* Retrieved November 2008, from http://curriculum.qcda.gov.uk/key-stages-3-and-4/organising-your-curriculum/principles_of_curriculum_design/index.aspx?return=/News-and-updates-listing/News/Teaching-of-new-secondary-curriculum-begins.aspx

Roberts, M. (2014). Powerful Knowledge and Geographical Education. *The Curriculum Journal, 25*(2), 187–209.

Roberts, R. (2013). *Powerful Knowledge: A Critique*. Debate Given at the UCL Institute of Education, 13 May 2013. Retrieved April 2019, from https://www.youtube.com/watch?v=DyGwbPmim7o

Robinson, M. (2013). *Trivium 21c: Preparing Young People for the Future with Lessons from the Past*. Wales: Independent Thinking Press.

Skelton, T., & Valentine, G. (1997). *Cool Places: Geographies of Youth Cultures*. London: Routledge.

Standish, A., & Cuthbert, A. S. (Eds.). (2018). *What Should Schools Teach? Disciplines, Subjects and the Pursuit of Truth*. London: UCL IOE Press.

Sturdy, G. (2018). Physics. In A. Standish & A. S. Cuthbert (Eds.), *What Should Schools Teach? Disciplines, Subjects and the Pursuit of Truth*. London: UCL IOE Press.

Vygotsky, L. (1962). *Thought and Language*. Cambridge, MA: Massachusetts Institute of Technology Press.

Wakefield, J. (2017). BBC News: Government Urged to Act Over Computer Science GCSE. Retrieved September 2018, from https://www.bbc.co.uk/news/technology-41928847

Ward, H. (2018). Curriculum Must Cover More Than 'Dead White Men'-Bousted. *Times Educational Supplement*. Online. Retrieved September 2018, from https://www.tes.com/news/curriculum-must-cover-more-dead-white-men-bousted

Young, M. (2008). *Bringing Knowledge Back In: From Social Constructivism to Social Realism in the Sociology of Education*. Abingdon: Routledge.

Young, M. (2011). Discussion to Part 3. In G. Butt (Ed.), *Geography Education and the Future*. London: Continuum.

Young, M. (2012). The Curriculum- 'An Entitlement to Powerful Knowledge': A Response to John White. Retrieved December 2018, from https://www.newvisionsforeducation.org.uk/about-the-group/home/2012/05/03/the-curriculum-%E2%80%98an-entitlement-to-powerful-knowledge%E2%80%99-a-response-to-john-white/

Young, M., & Lambert, D. (2014). *Knowledge and the Future School: Curriculum and Social Justice*. London: Bloomsbury.

Young, M., & Muller, J. (2010). Three Educational Scenarios for the Future: Lessons from the Sociology of Knowledge. *European Journal of Education, 45*(1), 11–27.

Young, R. (2013). *Powerful Knowledge*. A Debate Given at the UCL Institute of Education, 13 May 2013. Retrieved April 2019, from https://www.youtube.com/watch?v=r_S5Denaj-k

Young People's Geographies. (2009). The Young People's Geographies Project: Context and Rationale. Retrieved April 2019, from https://www.geography.org.uk/Young-Peoples-Geographies

Index

A
Academic disciplines, xvi, 2, 16, 42, 49, 50, 68, 75, 79–81, 83, 86, 92
Agency, 101, 112, 114, 119, 123, 148–150, 160, 164, 165
A Level content advisory boards (ALCABS), 7, 79, 80

B
Bernstein, Basil, 44, 81–84, 89, 155, 178

C
Capability/ies
　approach, 27–28, 100–102, 109, 139, 148–154, 159, 160, 163–165, 180–183
　approach to education, 26, 110–116, 125, 131, 155, 161, 183
　approach to education model, 112, 161
　deprivation, 101, 109, 112, 114, 118, 125, 153
　set, 100–103, 107–109, 112, 114, 148, 150, 153, 163–165, 179
Citizenship, 6, 10, 11, 23, 42, 77, 110, 111, 121, 150, 177
CIVITAS, 11
Climate change/warming, 1, 3, 11, 23, 45, 73, 77, 123, 124
Coastal geomorphology, 69, 85, 86, 124
Commission for Architecture and the Built Environment (CABE), organisation, 133

Index

Competencies, 2, 4, 8, 9, 33, 51, 74, 101, 116, 119, 180
Contentions (from research), 47, 103, 119, 120, 144–153
Curriculum
    classification, 81–83
    corruption, 38, 56
    extra/co curriculum, 33, 113, 184
    framing, 3, 44, 81–83, 169
    futures (see Future 1 (F1); Future 2 (F2); Future 3 (F3))
    hidden, 37, 38, 54, 114
    ideology (see Curriculum ideology)
    learnification of, 53
    making, 88–92, 118, 124, 134, 138, 139, 145, 147–150, 153, 159, 161–163, 165, 175, 180, 184, 186
    making workshop (in research), 143, 169
    methods, 36, 37
    national (see National curriculum)
    organisation, xvi, 154
    outcomes/secondary goods, 35, 39, 87, 117, 124, 148, 160, 163, 165, 182
    outputs/primary goods, 39, 40, 55, 110, 162, 163
    politicalisation, 38, 57, 80
    recontextualisation, 44, 55, 89, 178, 186
    structure, 5, 7, 8, 38, 39, 53, 83, 112, 180, 181
    thinking, xi, xii, xvii, 3, 26, 40, 54, 62, 67, 80, 86, 88, 89, 117–120, 131, 141–152, 155, 159–161, 165, 169–174, 180, 181, 184, 186
Curriculum ideology
    reconstructionist/radical, 42, 43
    vocational/utilitarian, 42, 115

D

Didactics (subject), 89
Doctoral research, xv, xvi, 27, 132, 139, 141–151, 153–156, 162, 169, 180

E

Education, 1, 34, 71, 101, 133, 160
    purpose, 11, 49, 151
Engaging places project, 132–133
English Baccalaureate, 13, 17
Epistemic
    access, 58, 186
    communities, 48, 50, 58, 175, 182
Eurogeo (organisation), 137
European Union Comenius funding, 137
Examination systems
    A Levels, 3, 7, 13, 50, 79, 84, 132, 150
    GCSE, 3, 7, 15, 17, 50, 79, 84, 154, 179

F

Fashionable causes, see Corruption
Finland, xvii, 21–23, 25, 27, 133, 137, 154, 187

Index   195

Firstspace, 132
Functioning, see Human functioning
Future 1 (F1), xi, 49–51, 79, 121, 138, 163, 166
Future 2 (F2), xi, 49, 52–57, 79, 111, 138, 165, 166
Future 3 (F3), xi, xii, 49, 57–58, 60–62, 67, 80–88, 99, 116, 118, 131, 138, 141–156, 159–188

G

GeoCapabilities
  in education, 4, 27, 120–125
  framework, 151–153
  project 1, 133–137
  project 2, 137–140
  project 3, 140
Geographical Association (organisation), 24, 79, 88, 89, 132, 137, 176
Geographical Information Systems (GIS), xviii, 22, 72
Geography
  academic, 70, 76, 176
  behavioural approaches, 75
  boundaries, 70, 73, 83
  discipline, 68–70, 73, 75, 86, 90, 123
  grammar of, 71
  great divide, 76, 81, 132–133
  human, 69–71
  humanistic approaches, 75
  key concepts, 71, 73, 87, 88
  language, 4, 84

  methodologies, 69, 70
  modern and postmodern approaches, 75, 76
  physical, 70
  process, 88, 90, 123, 150
  quantitative, 75, 76
  regional approach to, 75
  school subject, xv, xvi, 17, 67, 74, 75, 81, 100, 134, 160, 165
  skills, 72, 80
  thinking geographically, xi, 72, 73, 85, 123, 124
  vocabulary of, 71, 72
Globalisation, 1, 3, 23, 78
Graphicacy, 72

H

Habits of Mind (HoM), 8–9, 53, 54, 99, 147, 165, 180, 181, 184
Hirsch, E.D., x, 7, 50, 58, 78, 83
Human functioning, 26
Humanities, ix, 3, 10, 11, 24, 69, 81, 83, 87, 92, 111, 118–120, 166

I

Indoctrination, 55
Information computer technology (ICT), 21, 22, 177
International Geographic Union Charter for Geographic Education, 25

K

Knower structures, 48
Knowledge
  core, x, 7, 50, 78–80, 83
  essential, 6, 8–10, 12
  facts, 12, 80, 115
  horizontal structure, 83–85
  objectivist, 47
  of the powerful, 48, 51, 53, 60
  powerful knowledge, ix–xi, xvii, 3, 12, 22, 28, 57–62, 67, 83, 85–92, 99, 100, 116, 118–125, 131–133, 138, 140, 143, 145, 147, 149–155, 159, 160, 162, 164, 166–169, 172, 174–179, 181–184, 186, 188
  social constructivism, 47
  sociology of, 47
  traditional, 6
  turn (in education), 5, 7, 13, 14, 22, 79, 124, 174, 177
  vertical structure, 84

L

Lambert, David, xv, 4, 7, 10, 16, 17, 23, 24, 24n1, 26, 34, 50, 52, 55, 57–59, 71, 78, 80, 88–90, 120–125, 133, 137, 140, 143, 159, 166, 179, 181–183
League tables, xii, 8, 15, 39, 40, 114, 185
Learning power, 5, 53

M

Master's degree research, 187
Migration, 72, 123
Morally careless, 55, 57, 124, 150, 165, 173
Moral purpose, 11
Morgan, J., vii, 11, 17, 26, 55, 60, 78, 80, 88–90, 120, 121, 124, 125, 143, 181
Muller, J., xi, 47–49, 51, 52, 57, 79, 80, 91, 115, 159

N

Narrative (research), 144
National Curriculum
  big picture, 4–6, 22, 99, 147, 165–169
  changes, 9
National standards (US), 19, 137
Natural resources (teaching of), 17, 169–171, 174
Nussbaum, Martha, 3, 20, 26, 100, 101, 103, 105–107, 110, 111, 120, 121, 134, 160, 164, 183

O

Office for Standards in Education (Ofsted), 3, 16, 24, 74

P

Pedagogical adventure, 91
Pedagogy, 22, 33, 37, 38, 44, 49, 60, 88, 91, 92, 111, 113, 162, 185–186
  powerful pedagogy, 61
Personal, social and health education (PSHE), 6, 9, 10, 23
PISA ranking, 21
Post Graduate Certificate of Education (PGCE), 14
Powerful Geography (project), 20, 140

Index

Powerful pedagogies, 61, 62, 89, 90, 124, 149, 162
Professional development/INSET, 62, 137

Qualified teacher status, 16

Roberts, Margaret, 16, 60, 61, 86, 89, 124, 143, 149, 162, 175–177
Royal Geographical Society (organisation), 68, 79

School
    curriculum (*see* Curriculum)
    leaders/leadership, 2, 13, 28, 42, 111, 141, 151, 153, 155, 165, 179, 180, 185
    subjects, xv, xvi, 3, 9, 17, 67, 74, 80, 81, 88, 100, 115, 150, 160, 165, 178, 179
School ethos, 42, 161, 172
Secondspace, 132
Sen, Amartya, 3, 26, 99–120, 134, 152, 160, 164
Social realism, 48, 175, 181–183
Specialist/specialism
    subjects, ix, 13, 44, 153
    teachers, 13, 153
Structural features, 111–113, 117, 143, 147–149, 160–162, 172, 175

Teacher
    choice (in curriculum), 147, 153, 162
    professional role, 5, 14
    specialism (*see* Specialist/specialism)
    toolkit (publication), 172
Thirdspace, 76, 132
Transversal competencies (Finland), x, 21, 22
Trivium, 7–8, 178

United Nations (organisation), 25
United States of America, vii, xvii, 19–20, 23, 25, 133, 138, 154
University College London Institute of Education (organisation), viii, xv, 137, 140

Values
    non-public, 56, 114
    public, 56, 119, 161
Vignette, 85–88, 124, 169

Young, Michael, ix–xi, 2, 3, 9, 46–53, 57–60, 67, 68, 79, 80, 91, 115, 149, 159, 174–176, 178, 181–184